Jame

Surviving THE Meltdown

Staying Alive in 2000 AD

DLP Publishing

DLP Publishing
Wordsworth House, Ticehurst, East Sussex
United Kingdom, TN5 7HA

Published 1999 by DLP Publishing

A Catalogue Record for this book is available
from the British Library

ISBN 0 9535338 0 8

Typesetting and cover design by
John Eldridge Design Associates Ltd.
Sanderstead, Surrey CR2 9BX

Printed in Great Britain by
Cox & Wyman, Reading

To Elizabeth

Thanks for your support and
input all the way. For listening to
the arguments, for discussing the
scenarios and especially for
providing the insight to help find
a way through.

UH...DID ANYONE REMEMBER TO BRING A TORCH?

Surviving the Meltdown

How well would you cope if you had no electricity or gas for a month?

And you found the shops had no food?

Then somebody was ill in the house and you couldn't phone for help?

The Millennium Bug is threatening to make all this happen, and a lot more, but for a long time nobody has said very much about it. Government Ministers and senior industrialists have gambled that their 'millennium compliance' programmes will make it in time.

Now that the clock is ticking into the last months of the 20th Century, they are starting to realise that this is one huge gamble that looks increasingly as though it is not going to pay off. Masses of companies are having to accept the fact that they left it too late.

Nobody knows how badly it will hit, but power companies, utilities, food manufacturers and even airlines all around the world are admitting that they can't guarantee supplies

In Canada and the UK, police leave for December 1999 has been cancelled and many countries have put their armed forces on alert in case of civil disruption.

Hopefully it won't be Armageddon, but we'll still be in the middle of winter, and could face some very serious shortages as well as a lot of glitches in day to day life.

So far, every book on the Millennium Bug has been about the effects on business.

This one looks how it is going to affect real people. People like you.

And how you can get around the worst of the problems.

Our aim is to give you enough information and ideas that you make it through January 2000 healthy, fed, warm, and without losing all your savings or your sense of humour.

We hope!

What the press is saying

It's a well-known fact that all the best stories do not impinge on the public consciousness until the media gets hold of them. If you don't believe me, just ask Bill Clinton. The computer press has been aware of the Millennium Bug problem for ages, but it is only now beginning to hit the general media.

Just to give you a flavour of what some people pretty close to the problem are saying, here's a selection of headlines culled just before this book was printed.

MASSIVE NEW YEAR'S DEPLOYMENT OF CANADIAN TROOPS PLANNED
(Source THE OTTAWA CITIZEN)

BRITISH POLICE PREPARE FOR MILLENNIUM BURGLARS
(Source: The Times (UK)

RUSSIAN MILITARY ADMITS MILLENNIUM BUG IS A PROBLEM
(Source: Reuters)

WHITE HOUSE FEARS Y2K PANIC
(Source: WIRED NEWS)

PACIFIC GAS AND ELECTRIC WARNS CUSTOMERS OF POSSIBLE Y2K SERVICE INTERRUPTIONS
(Source: SPOTLIGHT)

CLINICAL EQUIPMENT FAILS Y2K TESTS DESPITE VENDOR GUARANTEES
(Source: YEAR 2000 INFORMATION CENTER)

IBM SEES Y2K HURDLES
(Source: Reuters)

BRITISH GOVERNMENT ACCUSED OF Y2K COVER UP
(Source: VNUNET)

Y2K FAILURE SHUTS DOWN U.S. SENATE'S COMPUTER
(Source: Roll Call)

MILLENNIUM EXPERT FEARS JAPAN'S DOOMSDAY
(Source: Reuters)

RUSSIA SEES TRANSPORT PROBLEMS FROM MILLENNIUM BUG
(Source: Reuters)

Y2K THREATENS U.S. POSTAL SERVICE
(Source: FCW)

AUSTRALIAN HOSPITALS FAIL Y2K TEST
(Source: THE AUSTRALIAN)

PROFIT WARNING TRIGGERS STOCK MELTDOWN
(Source: THE GLOBE AND MAIL)

CHINA FAR FROM SOLVING Y2K BUG
(Source: AP)

NEBRASKA WATER TREATMENT PLANTS FAIL Y2K TESTS
(Source: TECHWEB,)

BUG BITES IN UK
(Source: SUNDAY TIMES)

NEW YORK CITY'S 911 SYSTEM FAILS
(Source: CNN/AP, 1/1/1999)

STATE DEPARTMENT WARNS TRAVELLING AMERICANS ABOUT Y2K RISKS
(Source: Reuters)

1

Surely it's not going to affect me?

You may well be right. At least today.

On January 1st 2000 you may find it really *is* your problem.

Most people think of the Millennium Bug as being a corporate problem that will affect computers' ability to read dates. They don't need or want to understand any more about computers, and frankly I agree with them.

However, the whole of society depends totally on computers and their ability to keep working.

You may have no contact with a computer in your everyday life, or you may use one at work. Whether or not you realise it, your life depends totally upon them. Every time you start your car, a computer takes over and runs the engine. Every time you buy sfood or drink in a supermarket you're taking delivery of something that would not have been on the shelves without a string of computers.

Hospital life support systems, traffic lights, air traffic control, even telephones are all now computerised.

The electricity that comes out of your wall, the water from your taps, the programmes you watch on TV; all use computers to manage their supply and distribution.

They are also all totally interdependent, as you'll find out through this book. If just one computer goes down, say in a power station and the electricity supply goes off, everything in the region will grind to a halt. If the computer in the warehouse serving your supermarket goes down for a day, there will be 24 hours of no deliveries to the store.

Imagine if the bank's computer goes down – no cash in the ATM, no statements, and your bank account frozen.

This is no media-hyped doomsday scenario. It is a very real threat to the future of our lifestyle. Most people's reaction is that 'it couldn't happen.' Unfortunately there are far too many signs are that it will.

How the bug was born

Computers are now so intrinsically woven throughout our society, that people just do not believe that we could allow something like the Millennium Bug to happen. In fact we have sleep-walked our way into it.

When computers first hove into view on the horizon they were big, used lots of energy and had about as much memory as today's pocket calculators. From the outset everybody agreed they would have to use a date to manage all information, but because at the time memory was such an expensive commodity, some bright spark suggested just using the last two digits of the date for the year.

Instead of writing '1961' as a number, the programmers wrote the one and the nine as two text characters, with only the '61' part being a calculation. It means that from a computer's perspective 99 years is the limit of eternity, so when it reaches the end of year 99, its calendar will reset itself to zero.

In computer industry parlance, this is known as the 'Year 2000 Problem' which of course has been instantly abbreviated into the acronym 'Y2k.'

If you're having a hard time getting to grips with the concept, imagine what it's like for a computer, because computers are fundamentally stupid. Actually, in all

12

seriousness, computers have no intelligence at all. At the root of all their incredible complexity is a series of electronic switches that are either on or off. All of human life has been reduced to a string of ones and zeros, and if you don't believe that computers are stupid, try walking through an office building and listening to how many times you hear people shout 'Stupid Computer!'

So what will happen?

Different computers will react in different ways. Some will not recognise any programmes or information with a creation date that appears to be in the future. Others will simply lock up and stop working. Many more will just not even start up.

This is an incredibly widespread problem. The only computer manufacturer that has never suffered from it, because it has always used four digit years, is Apple.

The problem will also not hit all computers at the same time. Although some will appear to make it past the 'M' Day deadline, they may corrupt data that will not be needed until later in the year, perhaps at the financial year end in March, or even sooner by failing to recognise that 2000 is a leap year. Some may not show up until the end of the year, while others have already made their presence felt by miscalculating dates and charges for 1999. By the beginning of February, there were reports that over 50,000 Millennium Bug-related failures had already been experienced.

Although all computers have a date and time function, the display you see on the computer does not necessarily equal the same time as is being held by the computer's own internal clock. When you set the clock on a computer you do not change the 'real time' setting of the computer, only how it is displayed on your screen. The actual time the chip believes it to be is derived from the time that was imprinted on it when it was first created and programmed. The display in the corner of your screen comes from a programme that tells the computer to add or subtract a certain number of minutes, hours and days to its 'real' time.

This means that although your computer may well make it through into the new year, it could just as easily fail some months before or even after the actual millennium.

Some computers will appear to fly through the millennium crisis until the 29th of February 2000, when they will not recognise the day as being a leap year. Many of us might say 'so what?' but if the function managed by the computer depends on absolute accuracy, like paying the interest on your savings, then you would perhaps be more concerned. It also would mean that any work done that day on the computer might be lost as a result of the computer refusing to accept the date's existence.

So why don't they just fit a new chip?

If only it were that simple! It's not, because it's not just a chip problem. It's a software problem as well.

Let's start with the chips. Last year the industry sold approximately 7.5 billion chips. Just to get a feel of the true size of that number I'll write it out: 7,500,000,000. Fortunately these should, please note use of word 'should,' all be millennium compliant.

However, guaranteed millennium-compliant chips only started to make their way onto the market a few years back, so there are still an awful lot in use around the world that physically can't handle a four digit date.

According to industry estimates there are currently over 50,000,000,000 (50 billion) chips in use around the world. These can be found in just about every piece of electrical equipment you can possibly imagine. Even things that don't have a day and date display use them.

The Institution of Electrical Engineers publishes a list of the types of equipment that could be affected. The normal type headings are their words, the examples in italics are mine. Please note, I'm not saying any of these are going to fail, simply that they are examples of systems that could be affected by chip problems.

According to the IEE "Failures may occur in systems which:

(a) implement a timed control sequence or operate on a timed basis
Clocks, Washing machines, Pacemakers, Water treatment systems Flood barriers, Hydro-electric dams
(b) shut down unless a maintenance cycle is adhered to
Life support systems, Motor cars, Power stations, Manufacturing plants, Trucks, Airplanes
(c) produce regular reports
Power stations, Healthcare systems, Bank accounts, Stock market tickers, Medical monitoring systems
(d) report / handle timed events and alarms
Railway signalling, Traffic lights, Air traffic control, Security systems, Bank safes, Cookers, VCRs, CD players, Lifts.
(e) calculate totals over time
Bank account software, Mortgages, Pensions payments, Petrol pumps, On-board vehicle computers
(f) calculate averages, rates or trends
Investment Management software, Stock Exchange, Temperature monitoring, Safety valves
(g) rely on external timed data
Broadcasting and communications systems, Hazardous waste management, Chemical manufacturing, Animal feed dispensers
(h) rely on external geographical (GPS) data
Airplanes, Ships, Nuclear submarines, Satellites
(i) use or produce time-stamped data
Credit card readers, Clocking-in systems, Bank / finance transactions

(j) maintain historical state-of-system data"
Computers, Nuclear weapon systems, Chemical and biological warfare storage

That's a pretty exhaustive list, and it covers just about every facet of our society.

The most interesting thing about it is that in a lot of cases although they are managed by a chip with a day/date function, many don't even have a day/date display, Obvious examples include life support machines, cellular 'phones, and even cars.

Go into your supermarket and you'll see they do stock-takes by scanning bar codes with a hand-held unit that uses a chip, but you won't find a date display on the screen. Look inside the disc drive unit of your computer – you'll see lots of chips, but no display.

A lot of people ask why there should be a problem with a chip if it doesn't use the date function in its job. They are of course absolutely right. By and large these chips should carry on performing without a problem, unless they provide information to another system that relies on the accuracy of their date coding. There is also a very large number of chips that have been 'date-enabled' as part of their programming just in case they needed to deliver date-critical data at some later time.

Even if you only accept that at least ten percent of all chips do perform jobs that require a date code, or supply date-coded data, you still come up with a figure not a long way off ten billion chips that need to be identified, checked out individually and possibly replaced.

With less than 100,000 people in the computer industry with the skill to find them, the numbers work out at a staggering 500,000 chips that each person would have to find and replace. That's assuming of course that they know exactly where all these chips are to start with. Which of course they don't!

Even if they could find them all, they'd be stuck with the fact that there are hundreds of thousands of different variants of chip and software codes that have been used to program the chips.

We don't have a snowball's chance of finding all the chips that are faulty and repairing them before the clock strikes midnight on December 31. Everyone in the IT industry is just hoping that they've been able to cover the most important ones.

The software issue

Even when it's not a chip problem, there's the thorny issue of software non-compliance.

Over the last twenty years or so, computer software, the programmes that make the things run, has been written to work on machines that use two-digit years. It has only been relatively recently that software companies have developed four-digit year codes, and there are still billions of lines of computer code in use that rely on a two-digit year.

Faced with that problem there are two options. The first is to go through all the programmes in use, literally line-by-line, look for the code and re-write it. An additional provlem is that most of the old programmers who wrote the code have long since retired, and the programming languages they used have changed as well, so it's been left to a whole new generation of programmers trying to work through the code.

It's like giving a modern journalist the job of looking through the entire works of Shakespeare to find all the references to swords or knives.

There have been a lot of programmes designed and built purely to find the codes, or to test computers by emulating the effect of running in the new year. Unfortunately, some of these can cause more problems than they cure – One programme actually triggered a bug that caused the computers being tested to wipe all their data. A genuine case of the medicine being worse than the illness!

The other approach is to throw out all the existing computers and replace them with brand new ones. Globally companies have been throwing away old computer systems like there's no tomorrow.

Actually from a non-compliant computer's point of view that is the nub of the problem –all its yesterdays will become tomorrows. No wonder the poor things are confused.

The September problem

As if it was not bad enough for us to be facing a potential meltdown in our computer systems by the end of the year, there is another event worrying the minds of computer programmers.

It's called the '9 9 99' issue, or 'The September problem'

Although this is very much not a book about computers, to get to grips with this one you do need a little explanation, so please bear with me.

Everything that a computer does is controlled by programmes. These are stored on the computer's hard disk, and the computer's operating system, itself another programme, calls them off into the computer's memory. When you sit in front of your PC and type a document, you only see one programme running, but in fact there could be ten or twenty all running at the same time.

For example, let's say you are typing a letter to a friend.

When you hit the keys on your keyboard, a programme translates the action of depressing the key into a pulse that generates a character. Another programme displays it on the screen, yet another stores the character in memory so that the screen continues to display the text.

Now imagine that you want to print a copy of the letter.

You use the mouse, or the command keys on the keyboard, to tell the computer to print one copy of the document. This action calls up yet another programme, which on some computers will run along happily in the background while you get on with typing the rest of the letter.

These little programmes that carry out particular tasks would very soon fill up the computer's memory if they were not deleted from it once their tasks have been completed. Programmers use a variety of codes to tell the computer that the end of the programme has been reached and that it should now be deleted from memory. One of these codes is '9 9 9 9'

A large part of the millennium compliance effort has been aimed at tracking down any programmes or files that end with a series of '9's because there is a very serious concern among programmers that when a computer's operating system sees itself displaying this code, its reaction could be extremely serious. An added complication is that because the digits are written in a computer code, known as 'hexadecimal,' the same reaction could occur if the computer sees '9/9/99', or even '09/09/99.' In other words keying in the date on the ninth of September could be enough to halt some computers dead in their tracks.

At the moment nobody knows for certain what the effect will be, because although this code is popular, it is not universal. There are many computers that will not even be remotely affected, unless they rely on data from others that are. For those computers that do have a '9 9 9 9' problem options being foreseen range from the computer locking up or misinterpreting data, through to deleting all the files in its memory, including all its own programmes.

If the September problem proves to be half as bad as many in the industry think it might, then suddenly the whole issue of millennium compliance will leapfrog everybody's agenda and become the most talked-about issue in the news.

If it doesn't, then congratulations to the programmers for a Herculean effort. It would be nice, if foolishly optimistic, to hope that the whole Y2K problem could be settled just as successfully.

So what has the world been doing about it?

The last few years have been boom time for the computer industry, as companies spend gigabucks on new PCs, servers, printers and software. But very few companies can swallow the cost of replacing everything. According to several reports floating around the media, many smaller companies have realised that the cost of replacing all their computer systems is greater than the cost of going bust. There are going to be an awful lot of creditors looking for money from bankrupt clients in the new millennium.

Many of the world's largest companies have chosen to take a half and half approach – replace some parts of the infrastructure while tracking down and re-writing the weak links in others.

Even this approach is fraught with danger. The average large organisation uses so many computers and programmes that they have millions of lines of code that need to be analysed line-by-line to identify the date codes and rewrite the software to get around the problem.

Before they get to analysing the software, they have to carry out a millennium-compliance audit. In an ideal world this involves sending somebody around the company to seek out each computer, find out whether the computer itself is compliant, then check out all the software. It is incredibly time consuming to do this, and because it can only be done effectively by technically–competent professionals, it is very expensive. In most organisations the process has been further hampered by the fact that until very recently only a handful of companies kept strict records of what computers had been bought over the years and where they were being used in the organisation.

Too many companies started too late to have a chance of achieving the deadline of being millennium-compliant by 31 December 1999. At the beginning of January the sums on offer to millennium compliance consultants were running into five figures a day. One worrying aspect was reports of consultants working eighteen hour days to maximise their revenue. Great for them financially, but it's like asking somebody to read 'War and Peace' twice a day and keep alert throughout.

It's well known in the computer industry that large projects always over-run past the deadline. Unfortunately the deadline for this particular exercise is not movable.

The simple fact is that we're not going to make it.

Surely it can't be that bad – 'They' wouldn't let it happen

Although trying to achieve Millennium compliance has been the single most expensive exercise for society since the Second World War, more and more computer industry experts are becoming concerned about the level of quality behind the work.

During 1998 the UK government announced it would set up a task force of 'bug busters' and boasted of recruiting 20,000 people to spread the word among businesses. Unfortunately it looked from the outset like another case of too little, too late.

Let me give you an example that I came across through my regular work as a computer industry writer. Because I don't want to spend the last days of 1999 fighting lawyers I won't mention any names. The story concerned a company that had decided to implement a new millennium-compliant computer system. Before this, the company commissioned a 'Y2K' audit from a firm who did their work and presented the company with a detailed report on their computers and a summary of what they had to do to achieve compliance, followed no doubt by a substantial bill.

A couple of months later, while the new computer system was being installed, the engineers carrying out the installation found they were being asked to transfer a mass of programmes that had not been included on their list from the old system to the new. The users did not know whether or not the programmes were millennium-compliant, but the engineers suspected that many were not.

The list, from which these programmes were noticeably absent, had been created as part of the millennium compliance audit. After digging deeper to find out why these programmes were not on the list, the engineers discovered that the millennium compliance audit had been carried out largely by sending questionnaires to everyone in the company that used a computer, asking them to fill in details of their computers and programmes. In a frighteningly large number of cases the users had omitted very important programmes, either because they did not know what they were, or, because as one person said "I never use it, so I didn't think it counted."

Any one of these could have been the Achilles heel that hung the whole system up and stopped the company in its tracks. The results for the company's customers could have been devastating.

The sad thing this story points out, is that we don't know how many millennium-compliance audits have been done on the cheap through questionnaires filled in by people who, to be bluntly honest, don't know what programmes, extensions, macros, BIOS's and control panels are, let alone what they do.

It's a bit like you taking your car into the garage for a service and the mechanic handing you a spanner and a can of grease. Would you know what to do? Just to test the water I asked a friend of mine what programmes she has on her computer. 'Oh it's got Microsoft' was her only answer.

The only 100% accurate way to do a millennium compliance audit is for somebody to visit each computer in turn, list what is on it and lock it down so that nothing else can be added to it. The reason so many people have been tempted to cut corners is that the enormity of the task has been simply too great, as has the cost of hiring in people to do it. The task of bringing everybody's systems up to scratch has been described as being similar to sending a handful of people out to an Iowa cornfield and telling them to bring back all the worms by sunset.

As the year moves forward the newswires will be buzzing with stories warning about how companies are failing to make the cut. This report from the *Journal of Commerce Insurance News* at the end of 1998 gives a foretaste of what's to come:

A December survey of 110 U.S. corporations, 12 federal, state and local government agencies, and 12 industrial sectors has found that over half the respondents have already experienced Year 2000-related failures, and 98 percent expect more such failures in 1999.

At the end of 1998 UK government figures were suggesting that half of the country's businesses had still not developed a millennium strategy. At the same time the cost of hiring a millennium compliance computer engineer rocketed from hundreds of pounds a day to several thousand.

Despite all our best efforts, or at least those that everybody felt were economically justifiable, we're not going to make it. There is going to be disruption to our day-to-day lives, a lot of businesses are going to hit the wall, and nobody knows for certain how well or badly they are going to survive.

Every single aspect of life in our modern society is controlled either by a computer or some form of automated process. Early science fiction writers wrote about a society where people would have to do very little, but would have all their needs supplied by armies of robots. Mechanical men that would replicate every human function, but could never have a soul.

Although we don't have metal men, we do have factories where a handful of people control manufacturing lines that turn out products by the thousand. For the last ten years, commercial airliners have been built with no physical link between the pilots and the control surfaces, they 'fly by wire' relying on thousands of computer chips. Every time you go to a shop, your purchases are scanned over a barcode reader that sets off a computerised chain of events to ensure your purchase is re-stocked.

Even your car is computer-controlled – engine management is fully computerised which is why mechanics service cars by plugging in a diagnostic unit. The data downloads in many cars detail service history and events which of course are also based on date/time functions.

Everywhere you look in today's society you find computers controlling things. It would be a tragedy if mankind's greatest invention were also to be his undoing, but the Millennium Bug is looking more and more like the ultimate key to Pandora's Box.

Yes, But surely it will be fixed in time....

The biggest problem with the Millennium Bug is that nobody believes it is going to affect them personally. There seems to be a widespread belief that 'Somebody' will sort it out, or that 'They' wouldn't allow it to happen.

WRONG! The only person who will ensure you survive this is yourself, but then that's why you bought this book.

One of the most universally accepted truths about any project is that it will run over budget and over time. If you don't believe me just look at your own experience. Even if you've never worked in an organisation with a big new computer project, you can still find plenty of examples of this truism in your own life.

How often do you find yourself wondering where the year has gone and what exactly you managed to achieve? Or for that matter, how often do you set out to achieve something in the day, only to look at your watch some time later and be amazed at how long it has taken you?

Computer projects are just the same only on a very, very much bigger scale.

Back in the early 1990s when governments first started to become aware of the problem, they set up a series of deadlines for achieving millennium compliance. The entire computer industry all agreed that it really would be very much better for everyone if they could get the problem sorted by the end of 1997 giving a couple of years to test and re-test, and eliminate any bugs that might have been missed in the first pass.

Unfortunately, very few Chief Executives, Government Ministers or Finance Directors felt that it was worth committing massive funds to what they saw as a fairly minor software issue, especially as it lay quite some time in the future. Instead they preferred to take refuge in the belief that "somebody will sort it out."

This is a common mistake, psychiatrists call it the Willnott Syndrome. Broadly summed up it means an unshakeable belief that "it will not happen to me." If you want to see this demonstrated en masse every day go to California where over 40 million people live on the world's most geologically unstable phenomenon, the San Andreas fault. The police make a fortune out of the Willnott syndrome by catching people driving without a seatbelt, drinking and driving, and the good old faithful of speeding past a camera. They see its results all too graphically every time they attend a major road traffic accident.

For the first few years of the decade this mentality prevailed, leading to widespread apathy about the millennium problem. Meantime, the clock was ticking. Computer industry experts were voices in the wilderness, many finance people looked at them with suspicion and took the view that if their whizz-bang computers were all going to fall over in a few years time, then they must have been sold a pup.

The thought of having to pay money to sort them out rankled with the companies so badly that they did what all big businesses do when confronted with a corporate cock-up – they reached for their lawyers!1986 and 1987 were characterised by articles in the computer and financial press debating the legal liabilities, when they should have been waking the world up to what in the US was becoming known as the 'Millennium Bomb.'

Then, suddenly in 1998, just when the industry experts had said it should be all over bar the final testing, the world suddenly realised that this was a global problem that was not simply going to go away, so rather than argue about who actually struck the match while Rome burned, somebody had better call the Fire Brigade.

Even now, you will find people in the computer industry who still say that to have had any chance at all of leaving enough time to check things out thoroughly, everything should have been completed by the end of 1998. That would have left the whole of 1999 to test and re-test and eliminate any glitches that only came out of the woodwork once the new 'compliant' systems were running.

To give you a simple example of why they have to go back so many times try this simple exercise. Trust me, it's a perfect analogy.

Put a new bag in your vacuum cleaner and vacuum your carpets. Now take the bag out and put in another one. Now vacuum your carpets again. See, you've picked up at least half as much as you did first time, even though you thought the carpet was clean.

In Summer 1998, it was reported that the completion date for two of the biggest tax collecting departments in the UK was being put back from October 1998 to March 1999. As March approached they went remarkably quiet on the subject of how well compliance was being achieved.

Here's a couple of press cutting from the end of 1998 that sums it up all rather well.

LARGE UK FIRMS FALL BEHIND IN Y2K PREPARATIONS
(Source: COMPUTER WEEKLY (UK), January 15 1999)

In contrast with the conventional wisdom that big businesses are well ahead in terms Y2K preparations, a recent survey by legal firm DIBB LUPTON ALSOP has cast doubt over the readiness of large UK companies to tackle the Millennium Bug. Half the firms surveyed had spent less than 40 percent of their Y2K budgets; 69 percent had not even fixed their central computer systems by the end of 1998, and many of the respondents had yet to complete a systems inventory.

THE BIGGEST MISSED DEADLINE IN HISTORY?
(Source: COMPUTER WEEKLY (US), January 14 99)

The largely-missed December/31/1998 deadline, now almost forgotten by the world, is being called, "the biggest missed deadline in history," by Chris Webster, Head of Year 2000 Services at research firm CAP GEMINI.

President Clinton has now set March, 1999 as the new deadline for all federal agencies to complete their Y2K repairs, leaving just nine months for testing and implementation. When March comes, the deadline will no doubt be moved to June; when June comes, the deadline will be moved to September. But when 1/1/2000 arrives, says Y2KNEWSWIRE, all the missed deadlines and positive spin in the world will have no value whatsoever.

One more thought to cheer your up before we leave this section – You will keep hearing reports through the year about how companies are checking their systems out in order to prove they are millennium compliant. What they don't tell you is that they can't check out the whole system, because to do that would mean they have to switch everything off in order to make sure that the whole thing will work together.

They can't switch everything off, because that would bring their whole operation grinding to a halt.

It also appears from some press reports that as the deadline moves closer, companies are finding that the time to check everything out is being sacrificed in an attempt to maximise the time spent fixing the problem. Anybody who has ever hooked up a computer system will tell you that even if all the parts work perfectly on their own, it can take ages to try and get them all working together. There is a deepening despondency among the people who are doing the fixing that all is going to be far from 'alright on the night.'

Why did nobody tell us sooner?

Firstly because for a very long time nobody in authority believed it would be as bad as it is going to be, and then by the time they did they were between a rock and a hard place. In January 1999, A leading member of the British Government's Action 2000 campaign was quoted as saying,

"All our efforts to scare or energise people into action have failed. The dilemma now is how to raise anxiety levels without creating public panic."

Newspapers and magazines, are notorious for being immediate in what they cover, but as 1999 started to ease itself into being, some were starting to wake up to the issue.

By the time you read this the broadcast media will have cottoned on as well, but the vast majority of people, and here I'm talking in the upper 90%, still see the whole thing as some sort of business computer problem with very little personal relevance to themselves.

When I was first piloting the idea of this book to magazine editors I had some very interesting conversations. These guys are all intelligent people doing demanding jobs, and can be relied on to have a pretty good grasp of what's going on in the world. I was amazed at how many of them said something like "Oh it's a computer issue, I don't see how it's going to affect me. We have people to deal with that sort of thing in the company."

The most amazing reactions occurred after they'd seen the synopsis of the book. Suddenly they realised that our whole society is so dependent on computers, and so many are not going to be fixed, that it cannot but have an effect.

Now that there is a much wider awareness of the problem, at least among government and industry chiefs, the publicity machine has swung into disaster recovery mode. The last thing that Bill Clinton, Tony Blair, Oskar Lafontaine or the assorted global captains of industry want is a hugely unsettled population, so they are turning on the soundbytes in public while scampering around like startled rabbits behind the scenes.

The fact is that the Millennium Bug is going to have a serious effect on the world, but nobody knows precisely how bad it will be. Bill Clinton did hint at it in his 1999 State of The Union address. With masterful understatement he said "This is a big, big problem."

The more you read about it the more you feel a shiver run up your spine. Even if the bug itself doesn't throw the world into total chaos, the population's sudden realisation that it is going to lead to severe disruption is bound to cause all sorts of problems, especially as it is suddenly going to impinge on the world's consciousness round about the end of the year. By this time it will be far too late for any effective action. Senator Robert Bennett, the head of the US Senate's Year 2000 committee is on the record saying. "Even if the Y2K problem is solved, the panic side could end up hurting us badly."

As somebody that makes a living out of the media I am deeply committed to its purpose, Almost as deeply as I am cynical about it! The fact is that bad news sells papers and attracts TV viewers, so the media will do its utmost as 2000 approaches to hype the Millennium Bug. After all Armageddon makes for good television.

A word of advice is not to rely on media stories as your sole source of information. Certainly don't take them at face value. We have moved very much into a soundbyte society, so the skill at coming across well on TV is to deliver a snappy quotable quote. Certainly I would strongly advise you to treat with a huge pinch of cynicism any statement that uses any of the following phrases:

"We are undertaking an extensive programme and have every confidence…."

"You know, a lot of people are trying to whip this up…."

"We have put some of the best brains in the business to work on this…."

"You know, I don't think that's the real issue we should be looking at….."

In fact the millennium bug has developed its very own language of obfuscation. Saying your system is 'millennium-ready' sounds very comforting, but does not equal 'millennium-compliant.' There are a whole bunch of these trite expressions. For instance what exactly does "We are well along" mean? Could it be: "We know there's a potential problem and our computer people are working on it, but in all probability we don't have a cat's chance of making it.'"?

Other soundbytes to view with caution include:

"Taking steps"

"A plan is in place"

"Plans are underway"

and that all time ego-booster:

"Reasonably confident"

Will it be that bad?

Some people like to speculate that 'it won't be that bad,' but a lot of evidence points to the opposite view.

A friend of mine runs a computer staff agency hiring out people on contract work. He has a team of over 300 millennium specialists working for him, currently being hired out at more per day than you or I could dream of spending. He was asked how badly will it hit – "One in a thousand? One in a hundred?" His reply was cryptic. "It's going to be a lot more than one percent. My guys were being charged out at a couple of thousand a day in January '99. You can't imagine what we can get for them now. The only way to describe the situation is pure panic."

He also told me of an amusing counterpoint that perfectly illustrates the dichotomy. His company is constantly being pressurised by one large organisation to supply more millennium compliance engineers, despite their hefty price tag. According to some of the engineers, this particular company has left it far too late, and if it does make the deadline it will be more likely as a result of divine intervention than of careful planning and preparation.

Your sense of irony might be amused by that company's public statement on millennium compliance

"Together with one of the most respected firms in the computer industry we have been working steadily on a programme to ensure that when the end of the century is reached we will be able to operate all our systems perfectly. We have every confidence that none of our customers will experience any disruption as a result of the so called 'Millennium Bug' affecting us."

Weasel words? I seem to recall a similar sentiment expressed by the owners of the Titanic!

Why we can't just cut back to people doing everything

It's a lovely idea to think that when times get hard we can just go back to the old ways and we'll all muddle through. Unfortunately it is a non-starter.

The post-Second World War industrialised world was characterised by large labour forces working with equipment that was basically unchanged from the Victorian era. The driving principle of most manufacturing and support equipment remained one man to one machine, but over the last forty years we have been moving very rapidly towards a living in a fully automated society.

Automation and de-skilling have permeated every aspect of the way we live. We take for granted technologies that our grandparents would have seen as pure magic, both in our lives at home and in the way we work. Today mechanised farms of hundreds of acres can be run with a staff of four or five. Factories are highly automated, with robotic production continually replicating far finer tolerances than even the most skilled craftsman. In many industries people now only do jobs that are too complex and variable to develop machines to perform, or so lowly as to not be worth the cost of employing one.

If you bought a car between the wars you would find a

small hole in the front bumper where you would put a handle in and crank the engine to get it going. Even when they were first fitted with starter motors, the crank hole remained just in case the battery went flat. Today no car manufacturer would even dream of supplying a starting handle. Nor do power stations, railway engines or factory machines. Nowadays, you don't even have to strike a match to light a gas cooker.

Because we've become used to reliable technology we now take it totally for granted. Because we all use technology, there's no demand for non-technological solutions. Because of that, many old skills have been allowed to die out.

Don't believe me? OK try this simple test to see how your life would be affected by a lack of technology:

How often do you actually prepare a meal from raw ingredients rather than pull something out of the freezer?

Do you wind up your alarm clock at night?

Do you cook with gas, oil or electricity, or do you have a solid-fuel stove that you chop wood for?

Do you change channels by remote control or use a manual switch on the TV? Come to think of it, does your TV even have a manual switch?

When's the last time you walked up three flights of stairs instead of taking the lift?

How often do you walk to places rather than travel by car?

How long would you last without your phone/ PC/e-mail/ mobile?

Do you pay for things in cash or credit card?

All of these activities rely on advanced microchip technology. They could all be affected, one way or another, by the Millennium Bug.

The Bug and your home

As you've just discovered, the Millennium Bug is going to affect a lot more than computer systems in companies. Later in the book we'll look at how it could affect the infrastructure of society, but before the real scary stuff, take a look around your home and see how many things are likely to be affected by the bug.

What you need to identify is anything that uses an embedded processor to function. Fortunately in most people's houses that does not include too many everyday items.

The first things to check out are those that have day and date functions. Obvious examples are video recorders, home video cameras, music systems and televisions (you may not see them, but you will probably find that there is a function somewhere that gives you a time and date display, perhaps downloading from a Ceefax / Teletext signal).

The way to find out if they are millennium compliant is to write down the serial numbers and either telephone the company that made them, or go back to the shop where you bought them. Action 2000's website has a list of manufacturers that supply the UK and their millennium compliance statements, but in many cases these do not give specific model details.

Home computers bought in the last few months should be millennium compliant. I say should because there were still non-compliant PCs being sold towards the end of 1998. Certainly if somebody offers you a second-hand PC with anything less than a Pentium processor, I would be tempted to check it out very carefully. Unless its an Apple Macintosh, in which case you don't need to worry at all.

Most things around the home are unlikely to be affected. Unless you have one of the most sophisticated central heating programmers, you should be OK, but if it offers you a facility to have different settings for weekdays and weekends, then a call to the manufacturer might well be in order.

I've yet to come across a washing machine that offers day and date settings, but again they do use embedded processors so if you're concerned then a call to the shop is the answer. The same applies to dishwashers, tumble dryers, microwaves, cookers, fridges and freezers.

If you have a smart telephone, then you should check that out with the telephone supplier. Smart phones are those that use chips – especially the portable models. Home security systems are a potential risk, especially if there is any chance that you might have extra stores or cash at home. Home security systems are one area hat you absolutely must have checked out by a professional. Make sure that the person you ask is a member of the appropriate professional association. You can find their details in Yellow Pages.

Luckily we never quite cottoned on to the idea of the fully automated home of the future where everything was run by computers. However we live in such an automated society that although the bug may not have much effect in your home, there is no doubt whatsoever that it will have an effect on your home and on the way you live.

In the rest of the book we look at how badly the bug is likely to affect all the things we take for granted in society, and what you can do to minimise the impact on your life.

2

Will we have electricity in the brave new Millennium?

The power behind society

One of the fundamental precepts behind any thinking on the millennium bug problem is that we will need continuity of electricity supply. All the millennium compliant systems in the world will be useless if the power goes down.

Today's power infrastructure is very different from that which existed twenty years ago. In the first place it employs a fraction of the people. Think back a couple of decades and you'll recall that power company vans were a common sight. The roads seemed to be up continually as they ran cables through them. Engineers regularly came out and checked the distribution boxes, sub-stations and wiring, and if they needed repair or maintenance, they got out their tools and did the job.

Power stations were controlled by men who monitored huge banks of dials that told them the demand and what power was being generated. If they were generating too much or too little, they would make the necessary adjustments and the power station would work on. They were served by steam-hauled trains pulled from coal yards where they were loaded by men with shovels. If demand rose or fell the number of men required would also do so in line with the power stations' need for fuel. It was very simple, and it worked pretty effectively, but it relied on a large pool of skilled people to keep it going.

Then in the late 1970s and early 1980s power companies were privatised and their new profit-conscious managers set out to slash the headcounts as far as possible. They achieved this by two methods. One was to bring in as many computers as possible to replace people, the other was to re-skill the jobs of those who were left.

This is a trend that has been replicated throughout the whole of the last fifteen years. If you think around your own circle of friends and family, it's a fair bet that you know somebody who found themselves surplus to their employer's requirements, even though they had a thorough knowledge of that company's business, and probably an instinctive ability to get things to work when they'd gone wrong.

It's a great shame that we've done that, because as you'll see throughout the book the one thing that we don't have enough of today is people who can cope. If you want to know how much has changed, telephone your local power company and ask them how many people they employ today compared to how many they had a decade ago.

Odds are it's less than half. That's all well and good when the computers are all up and running and the supplies have not been interrupted by bad weather or any one of a dozen other problems, but if the Millennium Bug does half of the damage that's being predicted, there simply won't be enough people with the knowledge to put things right. Even the power companies accept that.

While this chapter was being written I checked out the UK government's 'Action 2000' website. The report of the Deputy Director General of Electricity Supply makes interesting reading:

"In electricity the independent assessment is initially targeting the 20 or so major players in the electricity sector, the major generators, transmission and distribution companies. With about

half the assessment complete, findings are that these companies are well on their way. They have largely completed work to rectify critical systems but some tests to demonstrate full compliance remain outstanding."

Just as I finished downloading that quote I had a fascinating telephone call from a contact about another section of the book. He expressed his concern that the government's published figures could only guarantee that at the time we were speaking only half of electricity industry's computer systems were compliant, and that he believed as many as 2% never would be."

"That two percent really worries me," he confided, "because nobody knows quite where it is. I know for a fact that between your office and the nearest conventional power station is a network of maybe two hundred miles of cable. There's at least a couple of thousand sub-stations and twelve main grid distribution points. Any one of those could take out a huge chunk of the network, like last year when a single cable failed in Auckland, New Zealand, and the city lost power for six weeks."

That set me thinking, so I spoke to a couple of power companies and was surprised by their approach. As with most companies they have the prepared statement for public consumption, which says that they are well on the way to achieving standards, etc. etc..

To a journalist that's like showing El Toro the red tablecloth, so it's into investigative mode and get the PR contacts to speak off the record.

After promising faithfully not to reveal the person's name or employer, we had the most amazing conversation. Here is what was said "Actually we don't ever guarantee power, which is fortunate, because there's no way we're going to get everything up to scratch in time for the millennium. We've concentrated on the systems that control power delivery to homes and offices and we're just hoping that this works. Undoubtedly, when we get back after the new year we'll find computers and processors that don't work. I'm just hoping that if we have serious problems, the telephone network will also be down, so I won't have to spend the first days of the new year fighting off the press."

Worried? Well I certainly was, because the whole electricity industry is a fine mesh of inter-dependent activities. It's like a house of cards; if one falls down, the whole edifice could crumble.

Incidentally, according to the host of sites dedicated to the subject on the Internet, the same situation appears to hold true in other parts of the world.

An Oily problem

Even if the power stations, pylon lines, distribution points and sub-stations all function perfectly, the whole infrastructure will grind to a halt if the flow of oil that now supplies most power stations is impeded. Oil comes from the docks through a network of pipelines, all computer controlled, and travels the world on computer controlled ships. Once it reaches its destination it is offloaded to refineries where it is 'cracked' to provide the refined products that keep society moving. Things like petrol, heating and diesel oil.

While looking closely into this aspect of the bug, I received an e-mail from a contact in the States who has a friend in the oil refining industry. Apparently this character seriously believes there is no way the refineries will make it before Millennium day, because it would be physically impossible to uncover, reach and test all their embedded chips without taking the refineries apart.

That seems a plausible thought, and as many oil companies run a thin inventory, with less than two month's supply in storage, the suggestion that the refineries might pack up caused quite a shiver. If that did happen, we would run out of petrol and oil products within a few weeks. Then everything else would grind to a halt very, very quickly. Including the power stations.

I called a couple of major oil companies and asked them if they could confirm or deny the story. Both responded with positive statements, but several weeks later, I noticed a complete absence of confirming documents being either faxed or e-mailed to me.

The only power stations that have their own in-built energy supplies are the nuclear ones. We can only hope that if the power does start to get iffy, they will put themselves into fail safe mode and shut down. Certainly that should be a reasonable expectation for our own domestic power stations, although it seems that the US government is more than a little concerned about what might happen to some of the old nuclear power stations in the former Soviet Union.

So what can you do about it?

You have a range of options that depend entirely upon how bad you think the problem's going to be. At one extreme, is to go out and spend several thousand pounds on a diesel generator that wires into your domestic electricity supply and will switch itself on automatically if your mains power fails.

The other is to buy a lot of candles!

The first thing you should do is contact your electricity provider and ask them to answer the questions set out on the next page. You will of course receive some reassuring answers, especially to the first couple, but by the time you get to the end of the list you may actually get the sort of information on which you can base a sensible decision.

1) How much of your infrastructure is now millennium compliant?

2) How much will be compliant by the end of October this year? (that allows them two months for testing, laughably short I know, but we're in their hands)

3) If you cannot deliver electricity as a result of the millennium bug, what are your contingency plans?

4) Is my electricity meter millennium compliant?

5) If you cannot deliver electricity to my house will I get a refund on my charges?

The last one's just for a laugh, but you never know, it might be worth a tenner!

You might be surprised to think that your electricity meter may not be millennium compliant. It depends on what sort of meter it is. Old-style meters should be, but many of the more modern types, especially the ones that have a capability of being read remotely, do use embedded processors.

You need to be particularly aware of the problem if you have a 'key card' meter, because the bug might not only affect your meter, but the equipment that 'charges' it when you buy more electricity on it.

It would be an awful shame if the power grid stayed up, only for you to find that your meter cuts off the supply. Do also bear in mind that if your meter is not millennium compliant so will be those used by a lot of other people and if there is a wide demand for replacement meters, you could wait a very long time for a new one.

One amusing idea that drew my attention was to send a form to your local power supply company, or indeed anyone that you depend on for supplies, and ask them to complete it and return it to you.

It goes like this:

To: *supplier's name here*
Please tick the box that best describes your current millennium-compliance position

☐ Our programme is well underway and we have every confidence that all our systems will be fully millennium compliant in time for 1/1/2000

☐ We have instituted a prioritisation programme to ensure that even if not all our systems are millennium compliant, none of those that affect our customers will be disrupted and we plan to maintain a full service

☐ We turned all our computers off last week and everything has kept going just fine!

Anything other than box 3 means you trust the reply at your own risk!

Seriously though folks, doesn't that just illustrate perfectly why you can't trust weasel words? Apart from the third option there's not a single statement on there that actually gives any firm undertaking that the supply from that company will continue!

So how do you survive with no electricity?

Certainly, if you have any concerns about your domestic electricity supply, and let's face it who doesn't?, it would be extremely sensible to make sure that every member of the family has a torch with at least one set of spare batteries, and that you have a battery or clockwork transistor radio in the house to catch news of what is happening in your immediate area.

Light is an obvious need, and although you may be

tempted to rush out and clear your local store of candles, remember that they don't give out much light, they can be knocked over very easily and present a fire risk. Camping lamps using gas canisters are a very good alternative, and you can find plenty of shops still selling kerosene lamps. If you can't find them in your local shops, look on the Internet or in magazines for walkers and campers. There are plenty of stores that could mail order some to you.

Whichever source of light you go for, make sure you have at least three times as many matches as you think you'll need. You will be amazed at how quickly you get through them. Butane cigarette lighters are another good thing to have around, but don't forget to have some gas refills as well.

We'll talk about food later in the book, but as we're covering power in this section, the one piece of advice to be given here is don't rely on your freezer. Not because it is unlikely to be millennium compliant, but simply because if the power goes down for more than a few hours, all the food inside it will go off.

If the power is out, remember the hospitals are going to be badly over-stretched, so you don't want to add to the general confusion by giving yourself food poisoning. Using food from freezers that has been allowed to thaw, can lead to all sorts of bugs. The worst thing you can do is to re-freeze frozen food that has thawed out, especially ice cream and dairy products. If you have any suspicions at all, better to throw the stuff away and not take the risk.

If you depend on electricity to heat your home, then you could be in for a chilly time. Remember the millennium bug will hit the world at midnight and half the world will be in the middle of winter. Some people take consolation in the fact that the Australians will get it first, but at least they will have warm temperatures and won't be at risk of hypothermia.

Get a hot meal inside you

The Australians are famed for celebrating Christmas and New Year on the beach with a 'barbie.' If you don't already have one you could do a lot worse than treat yourself to a gas-fired barbecue. Not only will you be able to enjoy it in the summer, if the power goes off, you'll still be able to feed everybody with a hot meal after New Year. Remember one of the best ways to warm yourself up is from inside, especially for older people.

Just make sure you get a couple of spare gas bottles. You will also have the consolation that if the power doesn't go off and you don't need to use it, you won't have wasted too much money!

Having said that you'd better place your order soon. It seems that supplies are running out fast, so you can guarantee that by the end of the year they will be priced high.

If you do buy a gas-fired barbecue, if at all possible don't bring it indoors to cook on, because there is a major danger of gas explosion associated with bottled gas appliances if they are not used very carefully. Both Propane and Butane are heavier than air, so if the gas bottle allows any gas to seep out, perhaps through a valve that is not secured properly, or from a slightly open tap, the gas will sink to the lowest point and 'pool.'

The slightest spark can ignite it, and as many owners of boats can testify, if this happens the resulting explosion and fire is awesome. If you do decide to bring your barbecue indoors, then make sure that it stands in an area where there is a good through flow of fresh air. Whatever you do, don't leave it standing indoors when it is not in use, and disconnect the gas hoses from the bottle every time you have finished with it.

Keeping warm

An obvious idea is to burn an open fire, but before you re-commission an ancient fireplace, make sure it will work. The first step is to get the chimney cleaned. Now that may seem obvious to you, but believe me there are a lot of people who would merrily set a fire in a fireplace that hasn't been used for decades, and the first thing they'd know about the years of soot and birds nests in the chimney would be when it went up in flames as well!

You can test if your chimney works by burning a handful of newspapers in the fire grate. If the smoke hustles up the chimney, then it's worth having it cleaned. If not, then you may find it has been blocked up by some previous owner of the property. Whatever you do, don't assume that just because there is a fireplace, there is a chimney behind it – a lot of house builders have put fireplaces into houses built over the last thirty or so years just for their effect.

On the basis that you have a good workable fire, then get your logs in early. As with so many things there is going to be a rush towards the end of the year as everybody else realises that the millennium is coming and wakes up to the things you've already suspected or learnt. As a result, not only will logs become more expensive, as will coal, but you will find that instead of being sold seasoned wood that burns well, the suppliers will start eating into their supplies that are not yet ready for burning. You'll end up with fresh, sappy wood that will crackle and spit, and you'll be constantly watching out for flying sparks that could set fire to your carpet.

Logs are great for a fast flame, but coal tends to burn a lot more slowly and gives off a more sustained heat for a lot less volume. It also doesn't spit.

If you do decide to heat your home with logs or coal, then you must buy two simple devices. One is a smoke alarm, a must for any house with a real fire, and the other is a carbon monoxide detector. If for any reason your chimney is not drawing fully, you could find that your fuel does not burn 'clean' but releases carbon monoxide into the air. You can't see carbon monoxide, or smell it, but it is the biggest domestic killer of all.

Generating your own electricity

You might decide to invest in a low output generator. This could be useful to run for a few hours a day, perhaps to keep the freezer cold or to run the pump on your heating system, but don't try to run the whole house – unless you buy a very big one, it simply won't cope. It is also worth remembering that generators are neither cheap to buy or cheap to run. If you buy a petrol powered generator, you will be amazed at how much fuel it drinks. Unless you're happy to top it up every two or three hours, think carefully about the inconvenience factor.

A word of caution: Generators are already becoming scarce. In the US many items that you would normally expect to be available from stock were on a twelve week back order in January, and the lead times are growing.

The world generator market is dominated by Honda. This is not a plug for them, the simple fact is that they are extremely reliable, that's why they have a share of nearly 80% of the world market for portable generators.

They are more expensive, but worth the money. It is also a good idea to buy from a reputable supplier. Garden machinery shops are normally the best places, they tend to offer a better after-sales service. The last thing you want is to buy from an unknown manufacturer, or chap with a truckload of cheap generators, just because it costs a hundred pounds less, then find that when you need it the thing won't start. If it comes to that on a cold, damp New Year's day, you can be certain you won't be able to get your hands on the guy who sold it to you in a hurry.

You also have to be careful with generators and petrol – Static electricity builds up on petrol cans and can cause a spark that could ignite the whole set-up. If at all possible use metal cans. Army surplus stores still have a good supply of 25 litre Jerrycans which are ideal. Put them on the ground, touch a bare metal spot on the can with the nozzle of the petrol pump to discharge any static electricity, _then_ open the can and fill it.

If you do buy a generator, get a competent electrician to advise you on how to wire it in to the house supply. Most small generators only produce enough power to run a central heating pump, a freezer and a light or two, so your best bet may be to run extension cords to those appliances so that they can be plugged in to the generator. Whatever you do, act early and get professional advice

The other danger with generators is fumes. NEVER, NEVER, NEVER run a petrol or diesel powered generator inside the house or the garage, unless it is designed to be built in and uses an external exhaust system. The fume build up in a closed space could quite literally kill you.

Another thought about generators concerns security. If, and it is an if, we find that power does go out for a significant time, you might be the only person in your area that has a generator. They are certainly not the quietest of things, so if possible it would be sensible not to run it in the dead of night when there's not a lot of noise going on in the area. There are times when it is good not to be noticed!

A last word on the electric front. If the power goes down, you might be amazed to find that society re-discovers the fine art of conversation, among other 'after dark' pastimes. Just make sure that you keep an eye on 'preventative measures.' An old friend who was a maternity nurse used to say that she could almost pinpoint the dates of power cuts by the increase in births nine months later!

3

A Very Big Bang

Today there is a plethora of organisations from whom you can buy your gas and they have all been actively promoting their wares. In many parts of the country you can even buy your electricity and gas from the same company.

The gas industry is now much the same as the electricity business – a fine mesh of interconnected parts, but there are many more gas suppliers, which means a very much higher degree of fragmentation, and even more opportunities for things to go wrong.

Despite the apparent complexity of their marketing, they all actually supply the same gas and run it through the same pipelines. They achieve their revenue through metering that measures how much gas each individual company puts into the national gas grid, and how much is used by their consumers.

It is a very sophisticated system, that could not operate without computers and embedded chips. Even the meters that they have been supplying for the past few years are microprocessor-controlled. If you have such a gas meter, especially one that uses a key card or pay-as-you-go system, you would be well advised to get a statement from your gas supplier guaranteeing that it is millennium compliant, or you could find yourself cut off.

Because the gas distribution system is computer-controlled, there is a chance of interruption to supply. At the same time as we checked out the electricity supplies we picked up the following statement from Action 2000 about gas:

"The Office of Gas Supply has received initial results from their consulting engineers on the independent assessment they commissioned. This shows that companies are well advanced in completion of their millennium-compliance projects."

I have to say that statement didn't exactly fill me with confidence, especially as an item on the ten o clock news the night before featured one of the government's millennium task force bemoaning a general lack of progress.

The gas industry's biggest concern is what will happen if the gas stops flowing. Unfortunately, there's nothing you can do, except make sure everything is turned off and wait for it to come back on again. The people with the big problem in that scenario would be the gas providers

To understand why a drop in the gas supply is the industry's worst nightmare, you need to look at how gas gets from the gas fields to the nozzle on your cooker.

Natural gas comes from gas fields, mostly under the sea, and is piped to onshore processing plants then into the national pipeline network. Gas flows through the national network supply pipes at a very high pressure into a series of distribution stations. These regulate the pressure down to about 20 pounds per square inch and feed local networks of gas mains that run under the street. From the gas mains, pipes are run to the houses and industrial units of the local gas company's consumers.

The gas pipe that comes into your house runs firstly into a regulator that drops the pressure even further, so that when you turn on the cooker, you get a nice steady heat, not a full-blown flame-thrower.

All the way down the line there are fail-safes built into the system, so that if the pressure gets too high, the valves will shut, and the risk of over-pressure hitting your appliances is eliminated. Many of these valves are controlled by chips, so there are a lot of links in the gas pipeline chain of control that could lead to the supply being blocked, either because the chip won't open the valve, or because it won't regulate the pressure properly. Whichever cause, the outcome would be the same– the supply could be interrupted.

If the pumps stop, or the distribution valves close, then the pipes will lose pressure as the gas within them is used by consumers. The first thing you will know of any problem is when the burner on your cooker goes out, or the boiler won't light in the morning.

In line with the laws of physics, if the pressure goes down, the pipes will not empty to a vacuum, but air could be drawn in, so the gas inside them would no longer be pure. The net result of that will be that the gas company will close down the system from the point of failure forwards

That's when the fun starts. Because the safe combustion of gas depends on having a supply of pure gas coming out of the burners, any air in the pipes makes it extremely dangerous. To restore the system, the gas company has to go right back to the root of the problem and purge every pipe downstream of the fault. They certainly will not turn the gas back on until they know that the whole of the affected network has been restored to working fully, safely and reliably.

If the problem were to hit a distribution point, say at the end of your street, the gas company would send somebody out who would shut it down, then go to each house in the street, turn off the gas taps on the pipes into the house and ensure that every single gas appliance was turned off. Not surprisingly gas safety regulations require that this exercise is carried out by a qualified fitter, and the world is not exactly overflowing with qualified gas fitters, so finding them and getting them to complete the task would be a very time consuming process.

Once the fault was fixed, the fitters would go back to the distribution point and turn the supply back on. Then they would go to each house and purge the gas pipe running from the street into the house, then purge the pipe to the cooker, then purge the pipe to the boiler, then turn it all off again to maintain the pressure in that part of the gas main while they go next door and repeat the whole process.

You can imagine how long that would take just to cover your street. Now imagine if the supply failed to your whole town. Every building in the town would need to be visited and inspected thoroughly, both to check whether or not there was any gas to the property and to turn off all the appliances, and subsequently to purge the system and restore the gas.

We are talking a very long time here. Carrying out this exercise for a street could take a couple of days, for a whole town you would be measuring the delay in weeks.

If that scenario does happen, the people who normally cook with gas, could soon find they lose their appetite for sushi!

Checking control systems

For once, people who do not use mains gas, but have LPG tanks, are ahead of the game. Their supply is not dependent on the mains, but on regular deliveries from their local supplier who turns up with a truck and fills the gas tank in the garden. From the tank to the house is normally a fully mechanical system, so the risks of LPG installations being directly affected by the bug are minimal.

In line with oil systems they are of course entirely dependent upon the local supplier being able to get to you and fill your tank. In view of the possibility of demand driving up the price as the year goes on, we've fitted an extra tank and are keeping them both topped up.

Whether you use natural gas, or LPG, or oil, if you have a constant-burning cooker, such as an AGA, or a central heating boiler, you will find that the supply to the appliance is regulated by an electrically-operated control box. If this has a chip, you need to check it out.

You should also check out whether or not the programmer that runs your central heating is mechanical or electronic. There are some extremely sophisticated systems on the market that can set different heat settings for summer and winter. The theory is that they are programmed once, then you sit back and don't have to worry about re-setting your heating ever again.

Assuming of course that they are compliant! The people to ask about that are the ones that maintain your boiler and cooker.

At least with gas you know what it will do to you. The same is not necessarily true of water.

Keep the taps flowing?

The water industry has been complaining for years about the problems of maintaining an ageing infrastructure. Certainly in Europe many water systems were built in the Victorian era, and they have served us well, but some countries' water infrastructures are balanced on a precarious knife edge.

We use many times the volume of water that we did only a decade ago. Washing machines, dishwashers, power showers, daily baths, health clubs and saunas have all driven demand for fresh water far beyond the predictions of the people who built the systems.

The average four-person family can get through fifty gallons of water a day. A standard bath takes twenty gallons, while a power shower can romp through ten gallons in five minutes. Add in about half a gallon each for drinking, washing and running the tap to get it nice and cold for a drink, and you soon see how it mounts up.

We also generate much, much more sewage. As if the volume of waste from the previous paragraph was not enough, its nature has also changed. Today we think nothing of shoving everything down the sink, from detergents and bio-shampoos to tea leaves and bacon fat, with the result that today's sewers are full of a sludgy, slow-moving, glutinous mass.

Stricter environmental controls on pollution have meant that water companies have developed highly sophisticated systems to re-cycle water as far as possible. It is not often discussed in polite company that the water coming out of the tap in most cities has already made that journey, perhaps as many as three times before.

Maybe it is not surprising that bottled water has enjoyed such a dramatic sales growth in recent years!

As with all the other utilities, the water industry is now a highly automated business. Everything is pumped around the network of pipes and sewers, and computers control where it goes and what happens to it. How long sewage spends in a re-processing pit, before going through the sand beds and chemical filtration plants, before being returned to a reservoir is monitored precisely by a network of computers and automated chemical analysers.

When the water leaves the reservoirs it passes through freshwater treatment plants where chemicals are added to purify it and make it clear. This too is computer controlled.

If the computers foul up, you could find waste backing up and becoming a playground for all manner of deadly microbes, not to mention the rat population. Or the cycles might not last long enough, leading to the risk of the fresh water supply being contaminated by sewage.

You will probably also not be in the least surprised to hear that I found the same lack of quantifiable public comment on the water companies' millennium preparedness.

Make what you will of their silence.

So how do you ensure safe clean water to your home?

This has to be one of the easiest sections of the book. Whatever happens at midnight on December 31, subject to the September problem, you can rest assured that the water should run clean and clear up until then. Unless you are metered, you will also be able to lay in a very substantial stock of water, simply by filling containers from your kitchen tap.

The reason for specifying the kitchen tap is that it is the one tap in the average home that you can guarantee is fed from the mains. The rest of the cold taps in the house are probably supplied by the tank in the loft, which unless it is absolutely brand new, will have a layer of sediment, or possibly worse, that will mean the water it contains would not be suited for drinking.

Bottled water supplies are an option, but an expensive one, and if you decide to buy a charcoal filter jug, you must remember to change the filters regularly.

Boiling water for fifteen minutes has long been an acceptable way of purifying it, and you can buy water purification tablets from your local camping shop if you want to treat water from the tank.

If you think that there's a danger of being without water for a long time, then you can become your own water provider and collect rainwater. Certainly if the weather carries on like it did for the last months of winter 1998 / 99 then there should be no shortage of the stuff coming down from the heavens.

If you do decide to follow this route, bear in mind that

the gutters around your house will have many years of dirt and other deposits that could harbour a variety of not very nice germs. As a precaution it would be worth checking them out before the leaf fall of autumn.

If you do decide to feed your down-pipe into a barrel to collect rainwater, then it would be worth buying a proper water butt with a lid and a diverter kit from your local garden centre. Then at least you start with a clean receptacle for the water. Even if you don't have to use it for a millennium problem, it could come in handy when there's a water shortage in summer and you want to water your plants. You will also find that proper water butts come with a tap that sits a little way above the bottom. This is to stop the water coming out from being contaminated by sediment.

A very handy hint if you do set up a rain water collection butt is to put it on a couple of blocks or some sort of stand. This will allow a decent clearance between the tap and the ground, so you won't find yourself in the embarrassing position of having sixty gallons of water but not enough room to get the kettle to it!

Do bear in mind that rainwater is not actually pure to start with. As it floats around in the sky it picks up dust particles and chemicals from the atmosphere so boiling or treating it with purifying tablets is absolutely vital.

Never drink any surface water, whether it has come from your own collection system or out of a river, lake or stream, that has not been either filtered to remove bugs, some of which can cause crippling diarrhoea, or boiled for at least fifteen minutes after straining. To strain water, stand it in a container to let the sediment drop to the bottom, then siphon water off the top through a plastic tube into another container through a filter made of a lady's stocking stuffed with layers of sand and charcoal.

Water can be treated to sterilise it with iodine, or with one to two drops of household bleach per pint. It can also be boiled, for at least ten minutes to make it suitable for drinking.

Boiled water tastes very flat, as does treated water. You can restore some of the flavour by oxygenating it. To do this fill a bottle about half full, screw the lid and shake it vigorously for a couple of minutes. This will improve the flavour dramatically.

Around the home you could try practising water conservation – certainly if there is a supply problem, you should make the best use of the water you have, especially when it comes to flushing toilets or washing. In countries where severe water shortages are normal, it is not unusual for people to wash with just a basin of water, which usually does more than one person, and then use this to flush the toilet.

The other alternative if you really want to justify blowing the budget entirely, is to have a swimming pool installed. Think about it – when else would you ever be able to come up with a reason for justifying such an extravagance!

Keeping communications open

The telecommunications industry is perhaps the one sector of modern life that can be forgiven for being upbeat about the Millennium Bug, because it has been spending fortunes in the last few years updating its systems to cope with the burgeoning demand for extra telephone numbers, fax lines and data links as the population takes to the Internet with a vengeance.

What the telecomms companies cannot do is to tell you for certain whether or not the particular telephone or fax machine that you use is millennium compliant. For that information you have ask the company that made it.

The effects on a telephone of not being compliant will be variable. Some will not display added value functions, like displaying a caller identity. Some will not face any problems at all, while some will not work. Only your manufacturer will be able to tell you how any one piece of equipment will react.

If there is going to be a significant weakness in the telecomms sector as a whole, it will be where it interfaces with other countries or possibly where it relies on satellites.

Satellites are a particular concern. Many of the older ones have been up there for some years and may well have a few non-compliant chips. If you doubt that the satellite systems are unreliable, ask somebody in the industry about what happened last spring when a communications satellite went off-line and millions of pagers died a sudden death.

Right now nobody is placing bets on whether or not all the world's communications satellites will come through unscathed, because nobody can tell for certain, and sending up an engineer to find out is a sheer impossibility.

If you are a fan of satellite broadcasting you might like to ask your service provider whether or not you will still be able to watch their programmes in the New Year. It seems unlikely that you won't, but there are some very old satellites up there that are still broadcasting TV and radio pictures.

Where on earth are we?

There is also a concern about Global Positioning System data. (GPS)

GPS is a system that uses a string of satellites around the world to provide navigational information. It works by the old geometric process of triangulating on fixed points and taking bearings and measuring distances. Ships' captains used to do it by measuring the angle of stars in the sky, GPS uses a much more sophisticated computer system to do largely the same thing.

Since the beginning of the 1990s when GPS was moved from being a purely military tool to being made available for civilian use it has enjoyed growing popularity. Civil airliners, merchant ships and cruise liners, even yachtsmen and hikers use it to locate their position on the earth. It offers astounding accuracy – some in-car navigation units can tell you their position to within three feet! It has also become a key component of many advanced security systems used in armoured security vehicles and expensive cars as a theft prevention measure.

The word got out a year or two ago, that the GPS calendar works on a fixed number of weeks, and that it is due to run out of weeks sometime in August 1999. What will happen then is that the almanac of weeks will reset itself to zero. Modern hand-held GPS units should be able to cope, but older ones may need a software upgrade, or might just not work at all.

We can have confidence that the airlines have got it right, but if you are a hiker or a small boat owner, you would be well advised to check out your unit with your local supplier.

There is also another small question that this throws up. If GPS goes down, apart from the chaos it will cause to mariners, is there a danger that the space agencies could lose track of the satellites and they would become just more lumps of space debris? Frankly I don't know, and I have not been able to find out, but it's a worrying thought.

Unfortunately stuff in Space has a nasty habit of obeying the laws of gravity. Small items make for spectacular light shows across the sky at night as they burn up on re-entry. However, the large ones don't burn up completely, and tend to make very big holes in the ground when they hit the Earth's surface. As you might have guessed by now, given that this seems to be the doom and gloom chapter, some of the satellites up there are quite chunky, so if any drop out of orbit we could be in for a few new craters.

Remember that one the Australians were so concerned

about a few years ago? For weeks the press was forecasting all sorts of dire consequences because it carried a nuclear generator on board, and it was scheduled to crash into the middle of the outback. As it was I think it landed in the sea, but I could be wrong. If you have any friends in the space business, I'd suggest you ask them for some more information.

So if you're not ready to take to the hills by now, with your backpack, bivouac and copy of the SAS survival guide, you must be feeling pretty confident about how you're going to survive the start of 2000.

Just wait till you read the next bit about Healthcare.

4

The Health of the Nation

One of the defining moments for many hospitals was the Christmas 1998 flu epidemic that swept through them putting an intolerable pressure on bed space and facilities. With an already over-stretched nursing staff, and never enough doctors available, the pressure of a national illness was too great and culminated in patients spending hours, in some cases over a day, waiting for care on a hospital trolley.

The health service is severely worried about the millennium bug. Strapped for cash, there are still many hospital trusts that do not have fully millennium compliant systems, nor the budgets to achieve them. The cost of millennium compliance consultants and engineers is now so high that to hire one of them for a day can equal the cost of carrying out two major surgical procedures.

Hospital managers argue with some conviction that the welfare of patients in the short term outweighs the considerations of mending computers for the future. Many health service managers are now focusing their efforts on developing contingency strategies to work around the loss of their computer systems.

From a patient's perspective this is severely bad news. Computers are playing an increasing role in the delivery of critical healthcare services. Life support systems rely on embedded chips, as do automatic drug dispensers that use electronically controlled measuring systems, heart monitors, oxygen level monitors, and EEG units.

Personally it would take a lot to convince me to go inside a Magnetic Resonance Image scanner or have an X-ray, especially if they use embedded processors. Unfortunately because you can't see the radiation, by the time you realise it's been harmful it will be too late.

The most disturbing aspect about embedded chips used in healthcare systems is that the hospitals are having to rely on manufacturer's undertakings as to whether or not the systems should be millennium compliant. This is achieved by the manufacturer asking the company that supplied the chip if it is millennium compliant.

If it is a brand new piece of equipment then they should have a reasonable chance of giving an authoritative answer. However, such assurances are not being so easily given if the equipment is old.

Hitting close to home

This chapter proved to be one of the most disturbing that I researched, because it brought home just how thin a thread holds society together when I started talking to Health Service managers. One in particular, who spoke to me only with a guarantee of total anonymity, told me about their preparations in some detail, and expressed his very deep concern on a number of issues.

Like many hospitals, his has introduced a very high level of automation and is doing its utmost to bring it all up to full millennium compliance. However, unlike so many publicly-uttered comfort statements, he was very dubious that they will make it.

Hospital computer systems are very complex. They use a lot of different programmes that all inter-relate with each other, so if you update one you usually need to update many others in order for them all to continue working together in harmony.

As part of the hospital's millennium compliance programme it booked in a series of computer systems suppliers to carry out their tasks in sequence. Unfortunately one had to cancel, thereby throwing out the whole sequence. Trying to fix dates to get them all back has been an absolute nightmare, as the companies are under increasing pressure from their other customers to give time to them.

In the end it is often the customer with the biggest chequebook that gets the most urgent service, so not only does the hospital face the prospect of a potentially non-compliant system in 2000, it might also be forced to work with a disrupted system this year as it is upgraded piecemeal when suppliers' consciences outweigh their other customers' incentives.

Although this was a worrying aspect of the problem, this particular manager's biggest concern lay not in the question of how well the hospital's computers would stand up to the millennium bug, but how well the hospital would survive the social disruption.

Although on the surface, the hospital's own resources of a huge generator, and a multi-fuel heating system, would appear to be able to overcome the loss of a major utility supply, such as gas or electricity, if either of these supplies were to be cut off, it could still turn very nasty.

The manager is absolutely convinced that there will be supply interruption in the New Year. He has been trying for some time to get his local energy suppliers to give him an accurate picture of their state of readiness, their plans for completing their millennium programmes, and their contingency plans should these be needed, but so far nobody has been able to convince him that he can count on 'business as usual.'

To cope with the expected demand, the hospital has

already cancelled all operations for the end of December and the whole of January. It has rehearsed its major incident plan and co-ordinated with the local police, fire service and the army.

"Let's face it," said the manager, "if the power goes down we're going to be the only building in town still lit up like a Christmas tree, so guess where everybody's going to congregate? We've got a lot of old folks around here, so all it would take is a couple of days with no gas or electricity and we'll be wheeling in the stretchers with more hypothermia cases than you could dream of."

He's also worried about the risks of social disruption. "Give it a couple of days, and a lot of people with nothing to do, and the drink will start to flow. We'll soon see punch ups, or worse, and we'll be the poor sods in the front line who have to deal with it. We'll also find a lot of people turn up here looking for food or warmth. Frankly, we'll need all the help we can get to keep them out."

His optimism about surviving without utility supplies was tempered by a summary of their position. "We have enough fuel to run the generator for fourteen days and to keep the heat going for about the same period. We are building up our drug supplies to keep us going for at least that time, but we can only store enough water for a day."

Seems to me that the last place to be at the beginning of January is a hospital, yet apparently many young ladies have decided that the 1st of January 2000 would be the perfect birthday, so have been scheduling their love lives to achieve a March / April conception. It's a lovely, romantic idea, but given that hospitals take it as read that they will have to deal with more than their fair share of partygoers and accident victims, regardless of what happens to their support infrastructures, I would have thought it better to wait a while

The Pacemaker problem

In a book like this there is a fine line to be tread between preparation and panic. The point that seems to be being missed by almost everybody in authority is that if you don't prepare properly than panic will undoubtedly ensue. Nowhere is this going to be more true than for people with heart pacemakers.

Pacemakers consist of a tiny box of circuitry that connects to the heart and makes it beat regularly. They can be re-programmed from outside the body using a special unit that downloads information from the pacemaker concerning service life and how it has been performing. If you contact pacemaker manufacturers they will tell you that almost all of their products are either millennium compliant or contain nothing that needs to be.

I have a friend with a pacemaker. Unfortunately he has had several, and suffered a lot of problems with them, to the point that his surgeon when fitting the last one told him that he would not be able to fit another with ease.

Last year he asked the surgeon if the current pacemaker is millennium compliant. Despite repeated requests, he has still not yet had confirmation that it is.

This is the ultimate two-edged sword– If you have an old pacemaker, do you demand a guarantee of compliance, for whatever that might be worth, or do you demand a new pacemaker with all the complications that major, elective surgery would bring.

And if you do demand a new one, who will pay for it?

Sensible preparations

Nobody can forecast whether or not they will fall ill, but there is a lot you can do to mitigate the dangers. One key point to remember is that if there is a post-millennium crisis, you will find that the emergency services will be stretched to breaking point, and that unless you have a really high priority call, you may have to wait a very long time for assistance. Even if your call is a life threatening condition, you might find that there are just not enough ambulances to get to you.

What makes the difference in 90% of cases of any trauma is the first few minutes of treatment. A trained first aider can quite literally save lives, so I would advise you very strongly to sign up and get the training. You never know who you might be able to help, and you'd feel dreadful if you find some day in the future that you could have helped somebody if only you'd known how.

Stock up your medicine cabinet. Not just with cold and flu remedies, but get in some decent bandages. It would be worth visiting your pharmacist with the checklist at the end of the book and asking if he or she has any other suggestions as to what you should have in stock. Then make sure you keep your supplies safe and secure and well out of the reach of children.

If you or any member of your family is on a prescription medicine, then go to your doctor and line up at least six month's supply that you can fall back on. There's a rumour going around the pharmaceutical industry that some supplies of drugs might be curtailed due to raw material shortages arising from either problems in the third world, or due to glitches in the supply chain.

If either of these happen, it could be a couple of months into next year before it becomes widely known, and you can bet your bottom dollar that panic will immediately ensue.

Even if you don't need to use your supply to cover an emergency, you will use it eventually anyway, so you will not be out of pocket. If you do have to use it you will be very glad you had it there when you needed it.

Lend a hand

Hospitals are very sensibly looking at the worst possible scenario and developing plans to manage it as best they can. Most have embarked upon an extensive programme of working with suppliers to try and ensure that they have enough drugs and food supplies to cover any potential disruption, but all agree that if there is a big problem with healthcare equipment failing they will desperately need more staff, and there are simply not enough nurses and doctors out there to cope with the volume of patients looked after by today's computerised health service.

If the worst comes to the worst, health service managers are going to need assistance from anybody who is prepared to turn up. Don't be afraid to volunteer.

Assuming you can get through the traffic jams.

5

Cars, Trains, Boats and Planes

After all the doom and gloom of the last few chapters it's time to lighten up a bit and look at things that might bother us, but should not unduly harm us. Which immediately brings to mind the transportation infrastructure.

If you go into the travel agents right now and ask about booking a holiday for the millennium, odds are that the counter clerk will smile sweetly at you and say "I think you've left it a bit late!"

She will probably be right. Millennia and tourism seem to go together like fish and chips. It is a little known fact that the start of both the first millennium and the second (the years 0000 and 1000) were also marked by huge numbers of people moving around the world. The first millennium (0000) found the world slap bang in the middle of the expansion of the Roman Empire, while the start of the Second found William the conqueror heading off to a beach in Hastings to take over Olde England.

Hopefully the next millennium will come into existence more peaceably, and I know the purists among you will say that the Norman Conquest happened in 1066, but remember that the years were all re-set when the Gregorian calendar was established in the 16th Century, and even in the Dark Ages the French were not famed for punctuality.

Towards the end of 1998 the 'ever-so-exotic places to

spend the millennium' articles started appearing in the press and people by the millions flocked to book them. "Be the first to bring in the New Year – Fly with us to Auckland" hailed one advertisement, thereby coming up with what has to be the single most original idea for getting tourists to visit one of the world's least sought out tourist destinations.

"Christmas in Bali, New Year in Hong Kong," proclaimed another. "The Cruise of a millennium – See it in twice," offered one enterprising ship owner, that plans to straddle the international date line.

They all sound extremely enticing, until you see the cost, but despite the massive hike in airfares, including one airline removing all refund options on its flights for two weeks either side of 31 December 99, millions have booked for what is expected to be the world's biggest party.

However, as you might expect, not one of them has given a single thought to the potentially tricky matter of getting home after the party's over.

The 'travelling home' problem

During the middle of 1997 when researching my very first article about the Millennium bug, I telephoned one of the world's largest airlines and asked about millennium compliance. The reply was very cheery "Oh we won't have a problem. Our planes will all be fully compliant. We'll be flying a full schedule."

So that was alright. Nothing to worry about there. I duly reported the conversation, along with a few other ideas, there was even some talk about a charter operator booking Concorde for a round the world supersonic flight that would somehow manage to fly through midnight in several different parts of the world, thereby allowing the passengers on board to celebrate the millennium as many times as they could wish for.

Wonderful stuff, and so I was happy to report that statement in further articles until about eighteen months later. Then, as 1999 drew near, it seemed appropriate to be a good journalist and check up on my facts. (Yes I accept that sounds like a strange concept for a journalist, but it has been known!)

Same airline, new story, same question. This time the response was far from chatty and went something like this:

"Whilst we have every confidence in our own ability to meet the demands of millennium compliance within both our aircraft and our operational systems, our primary concern is for the safety of our crew and passengers and we will act accordingly with respect to operations over the millennium period."

Off the record, the PR contact was much more enlightening. "Frankly, we have a real concern about air traffic control in some countries, and the company view is that if we can't be certain that a plane will land safely then it just will not leave the ground."

Seems reasonable, but then while talking to a pilot friend of mine another interesting aspect of the problem came to light. Apparently seventy percent of the world's airliners are normally in the sky. Many of the world's major airports have up to ninety percent of their 'stands' occupied for a large proportion of their working days, so they are not exactly overflowing with parking space for airplanes. In fact, if they all tried to land at major airports at once, you could fill up every plane sized parking place and still have half of them flying around in circles looking for somewhere to put down.

"So why not just rotate them around the world along with the midnight time change?" I hear you ask. Simple – the whole world's airline industry flies on Greenwich Mean Time, so when the problem hits, it will hit the whole lot at once.

Many of the world's airlines are very sensibly agreeing

that the sky is probably not a good place for their planes to be at midnight on 31 December, not that they have anything to fear of course, but some of the excuses they are giving out to the public are so clever they have to be heard to be believed. This statement is from one of my favourite airlines, and shows how you can say something that tells the truth and nothing but the truth, but singularly fails to tell the whole truth!

"We have always been an airline that responds to our customers desires, so we carried out research to find out what they really wanted from us around the time of the millennium. We found that the vast majority of our customers, and to be fair our crew, said that by the time midnight comes, they would much rather have reached their destination and be celebrating, than be in the air trying to get there. In line with our customers' wishes we have modified our flying schedules accordingly, and so it is fair to say that we are unlikely to have any planes in the air at the midnight hour."

You have to admire it don't you? I mean what a wonderful piece of word-smithing. It totally fails to reveal the actual reason why they are not flying, but instead presents a wholly plausible excuse that even manages to show you what a caring company they are.

How much of an excuse becomes clearer when you start to dig a little deeper. Again this information comes courtesy of my pilot friend.

So far, it appears that the Air Traffic Control business world-wide has been very reticent about guaranteeing its ability to meet the millennium deadline. That's why the airlines are playing safe. But whatever the public statements, there are a lot of countries to which airlines fly that they simply would not trust to get it right, however comforting their statements may be, and they have absolutely no desire to park their expensive airplanes anywhere that they think there could be a chance they might get stuck.

Apparently there is a huge demand in the airline industry for parking places in the US and mainland Europe. Military airfields are being looked at, as well as freight only airports. My pilot friend said it could come down to 'anywhere with a long enough strip of tarmac and access to a fuel bowser.' They have to be a bit careful – many of the landing strips being looked at are fine for getting a plane into, but not long enough to get it back out again.

This could cause a lot of problems, because airplanes are not like buses plying the same route day after day. That's why a delay on an airline's flight from London to Puerto Rico on Tuesday could throw a spanner in the works for its customers flying from London to Jerusalem on Thursday – If the flights use the same plane and it is 'out of position' then as much as a whole week's flights could be affected.

The way things are looking, it could be that much of the world's airline industry will have all its planes concentrated in a lot of unusual places. Not only could it take them quite a while to get them there, it would take just as long for them to get them back to position after the midnight deadline.

If you are travelling abroad, you'd better be prepared for delays. Remember, millennium bug or not, airplanes do go wrong anyway, and it won't take many to have a fault for the whole system to clog up.

On the other hand, you could stay home and get yourself a job as a security guard. What an unusual way to spend the millennium – babysitting an Airbus on a closed-off section of motorway!

Take to the Highway

Closed off motorways leads me nicely into the thought of what will happen to the road infrastructure in the New Year.

Whether or not there is a millennium problem, there will certainly be a lot of congestion as people travel to and from parties, and you can bet your life that the constabulary will be out in force to keep an eye on proceedings and rightly detaining anybody guilty of celebrating too much and getting behind the wheel.

They could also find themselves doing a lot of work just to direct the traffic, especially in view of the concerns expressed about traffic lights.

Traffic lights are yet another of those things that we all take for granted. Yet how often do you notice that you can be heading down a street and find they all turn red just as you come up to them? No, you're not jinxed, and yes it is deliberate.

Traffic light systems use a very complex computer programme that relies on network of sensors under the road to monitor the volume and flow of traffic. That's why major cities like London do not lapse into gridlock twenty four hours a day, despite a near doubling of car numbers in the last decade. In fact the systems keep the traffic moving very smoothly. The same sort of idea has recently been introduced on the M25 where speed limits are varied in accordance with traffic volumes – the greater the volume, the lower the speed is set so that traffic flows smoothly and doesn't bunch. Despite the motoring public's initial cynicism, it actually works very well.

You have probably noticed how one set of lights going out in a city, especially in rush hour, can lead to horrendous jams that sometimes feed back for two or three miles. Now imagine what would happen if all the lights in an area went down because of a computer problem, or if the chips went haywire and started sending in erroneous information.

The system should of course go to its fail safe position which is that all traffic lights would go red. As a result, the whole city would immediately grind to a halt. Sorting out that mess could take a very long time. If you've ever tried getting to a cup final, or a 50,000 seat pop concert by car, you'll know from your own experience how the traffic chaos can add two or three hours to your journey. Just imagine a whole city in that state. Maybe you'd better take a packed snack with you when you go off to see in the New Year!

The other, and possibly even more frightening scenario, would be if the processors controlling the lights got the sequence totally out of sync, and the lights at some junctions showed green to all the roads approaching them. It would not take long for the junctions to be turned into major car wrecks, and you can bet the ambulances and fire engines would not be able to get through, especially if as predicted, it all goes wrong in the wee small hours of the morning.

I did have a thought that if the book doesn't sell, I'd go out and buy a people carrier and get into the Millennium Minicab business. After seeing how much minicab drivers can get away with charging on an ordinary New Year, it was very tempting.

Then after researching the traffic light scenario I thought better of it. Even if the revellers were prepared to cough up for my exorbitant millennium night rates, it just would not be worth the hassle to get stuck in to all that traffic.

On the other hand I do reckon there will be a lively trade in rescuing motorists whose cars have been hit by chip failure.

Chip failure? In a car?

Oh yes, I'm afraid so.

Back in the mid nineteen-eighties, motor car manufacturers realised that the key to producing more efficient engines was to do away with manual fuel control systems whose concept had been largely unchanged since the turn of the century, and substitute computerised engine management units that could vary the mixtures and timings with much greater accuracy.

The first ones had the ability to vary the petrol/air mixture and advance or retard the timing of the engine, and soon showed significant contributions towards better fuel economy.

All very 'green' and laudable. Then the car makers realised that they had a whole lot of computer processing power available that they could put to much better use than just tweaking the engine, and they found all sorts of interesting things for it to do.

Nowadays, car computers monitor the whole vehicle. They tell you if a door is open or a bulb has blown. They manage the display of dials on the dashboard. They tell you if the car is due for a service, and along the way, they sample the engine's performance millions of times a minute to make sure it is running just right.

Sometimes they give you a very handy display that says something like Friday February 5.

But wait a minute. How does it know that February 5th is a Friday. When the end of February arrives and it says Monday March 1, you would be wise to ask how did it know that this February isn't a leap year? Answer – because it has a day/date chip.

These chips have been in just about every popular car for the last decade in one form or another. They are just as widely used in commercial vehicles, including the heavy trucks that deliver everything to shops.

I had a car like with a day/date display until a year ago when I finally got an answer from the manufacturer about whether or not it would switch accurately at the millennium. That was when I sold it. And no it wasn't very old, actually it was a1996 model year, and I was the first person to go into the dealership and ask the question.

Subsequently a poll of car manufacturers, including those whose cars I am currently interested in buying, has revealed that some are giving millennium compliance undertakings on all models, while others are only prepared to guarantee those made after a certain model year. In some cases this is a lot more recent than you might think.

What would be the effect?

Most people in the motor industry tend towards the view that a non-compliant engine management unit will encounter problems as soon as it wakes up when you turn on the ignition and it runs through its checks. If this occurs at the end of a superb party on millennium night, you could find you have a very long wait for the recovery man to get to you.

If the engine management unit has a service awareness function and a two digit calendar, it might decide that it has not been serviced for a very long time because it won't recognise the data that suggests its last service date is still some 99 odd years in the future.

At best it will throw up an 'error' message on the driver's display. At worst it will not start, especially if there is a safety consideration wired in the programme that won't allow it to run if its service record does not appear to be in order.

In the middle are all sorts of interesting possibilities. Certainly the day/date function might go awry, remember that Jan 1 2000 is a Saturday whereas in 1900 it was a weekday. You could find that the readouts that tell you how many miles to the gallon you get go haywire. You might even find it completely blanks its memory and forgets how many miles it has done altogether.

If the engine management unit is likely to do either of these things, then it will need to be reset by the garage. If possible you could get it re-programmed to think it is ten years younger, or if that is not possible, then a replacement engine management unit might be the only solution.

Engine management units are far from cheap, so if you have any concern about your car being capable of running after the millennium changes, take it to the garage. There will be a lot of people going through this exercise, especially after September, so you could be faced with a shortage of replacement units if you don't get in early.

The other option is to quietly trade in the car and buy a newer one that is millennium compliant. Certainly the used car trade is none too buoyant at the moment anyway, and later in the year its appetite for non-compliant cars will be non-existent. You could find yourself hearing that classic used car dealer's comment. 'One of them? Sorry 'guv' I can't even give 'em away!"

Act now. It could save you a lot of hassle later.

Sailing into stormy waters

Trains, planes, trucks, cars and boats all have something in common – a limited range between filling their tanks. Like so many other things in society if the fuel that keeps them going dries up, then it will make no difference at all how millennium compliant they are, they won't be going nowhere!

Fuel, in the form of crude oil, is moved around the world in huge tanker ships that take it from oilfield to refinery. Oil is almost always refined in the country it is to be sold in, as carting oil products around the place is a very expensive pastime, not to mention one that has more than its fair share of hazards.

From the refineries, the refined oil products, such as petrol, diesel fuel, lubricant grades and tar, make their way to their destinations either by trainload or road tanker. The world currently has a major over-supply of the stuff, so at least there is a lot of it around in storage tanks and on ships, but it does depend on the transport infrastructure to get to the supply points where you and I can buy it.

The ships that bring oil to our shores are, as with most things, highly computerised. Apart from using GPS systems for navigation, they are also heavily reliant on electronics both for auto-pilots and to relay commands from the bridge to the engine room and rudder.

The Marine industry as a whole has perhaps been one of the slowest to react to the threat of the bug. As late as last June, the shipping industry newspaper *Lloyd's List* was advising its readers of the dangers of the bug, sadly acknowledging the fact that "awareness of Year 2000 problems was low and certainly below par when compared with other industries."

This was a sad indictment of the state of affairs in the one industry that really does make the world go around. Whatever commodity you think of, whether it be oil from the Gulf, bananas from the Caribbean, raw materials for industry or finished goods from cars to computers, you soon realise that every item we use in modern life would not be with us, had not some, if not all, of its parts been moved around the world by ship at some time in their life.

Within the article, by a respected member of the ship operating industry, himself a former director of a major shipping fleet, the extent of the problem was summarised very clearly as being as much down to embedded chips as to anything else.

"A worrying survey by the US Coast Guard, after it declined to believe confident assertions of compliance from manufacturers, discovered 20% of the chips tested were in fact non-compliant. No wonder nobody will commit themselves."

With just as many obstacles ahead of them, and a lot less time to get to grips with the problems, it seems that ship owners are facing an almighty task. Hopefully it will not turn into a doomsday scenario but will probably manifest itself in a lot of minor aggravations that can be managed by a well trained crew. However, as with so many industries, shipping has been laying off people like you wouldn't believe.

No doubt any problems will be made even worse by the fact that many of the old sea-dogs have been replaced by much younger, less experienced crew, and ships are often crewed by unskilled deck hands hired in the cheapest countries. Some of the largest deep sea ships travel the oceans of the world with crews of ten or less. If one small problem strikes they would cope with it very easily. Two problems would start to put a strain on the crew. Three could be too much for them to cope with.

Shipping, like so many industries is also a business that is highly dependent on all its parts moving in harmony. Although the obvious areas to look at in terms of millennium compliance on-board ships concern command and control systems, there are plenty of other problems at the dockside that could surprise an unsuspecting ship owner.

Dockside movements are now very much in the hands of embedded processors that regulate all manner of functions from managing the containers to be loaded on or off the deck, to handling stores and 'housekeeping' items. Imagine the reaction of a jumbo jet load of passengers arriving for a Caribbean Cruise, if their ship could not let them board because a faulty dockside processor stops it from emptying the waste from its previous journey

That seems to me like an extremely good excuse for not undertaking any overseas trips around the end of the year, but for staying close to home and stocking up with petrol. As already mentioned with regard to generators, a few army issue Jerrycans could come in useful.

Unfortunately it also looks like I'll have to leave the gas guzzler at home and take the smallest car in the family. We may not be as comfortable, but at least we'll get a lot further.

So it looks like we'll be on the train then?

Well at least this is one are where you should be able to breathe a sigh of relief. I mean how complicated is a railway engine for goodness sake?

You guessed it – not very, but that doesn't mean you won't find a goodly handful of chips controlling everything from power flow to on-train heating.

Knowing the rail industry, safety will prevail, so any train that is not 100% functional will not depart from the sidings. But even if the train is safe, can we be sure that the network is? According to the company that runs the railway network, it plans to have 100% operations with all signals working and all power supplies intact.

Heard it all before? Probably so. But at least if the rail network does suffer glitches, the only concern will be that you find you can't take the train to work thereby getting a day or two more on your holiday. Oh, and that the fuel trains to the power stations might not get through. And… Need I say more? We're back in the realm of interdependencies again!

Actually in some parts you can travel by train with 100% confidence that they will not be affected by the millennium bug in any way shape or form. These are of course the preserved steam railways. Not a single micro-switch, thyristor or chip in sight, other than the greasy ones they sell with fish in the cafeteria, and all the signals are operated by pulling handles that move them mechanically.

The only problem is that even though they are among the most popular tourist attractions in the country, most of them go from nowhere, through nowhere, to nowhere.

Ah well, such is progress. You may not be able to go very far for a while, but at least you'll still have a job when the trains get back to normal.

Or will you?

6

Working next January

Even assuming that you can get to the office, despite the hold-ups forecast in the last chapter, the big question is not so much 'What will it be like coming back to work after the world's biggest party,' as 'Will we be coming back to work at all?'

That may sound like an awfully big, bold statement, but throw your mind back to the beginning of the book where we read comments from computer professionals and business advisers that any company not prepared for dealing with the bug by the beginning of 1999 had very little hope at all of surviving its effects.

At the turn of the year, the UK government's Action 2000 campaign launched a new advertisement aimed at businesses after discovering that "a staggering 51% of businesses employing between 10 and 249 staff had yet to take any formal action to tackle the Bug."The reason it has come to this, is that too many business people believe the Millennium Bug will not affect them and that someone else has the matter in hand.

Many of these firms use computers to give them a competitive advantage, and are totally dependent on suppliers for the continuation of their business, yet they are quietly sleepwalking their way towards the total annihilation of their enterprises.

Don't believe me? Ask anybody in the Disaster Recovery business and they will confirm from their own experience of events like terrorist bombings or massive computer failures, that over half of companies that lose their ability to work for five days find their customers go elsewhere. Companies that are out of business for ten days or more almost always find they never get back in.

The lack of compliance activity was even more concerning among companies with fewer than ten staff, where only one in four had taken any steps at all.

Everybody who wakes up to the problem asks "Who is going to be hit, and what can we do about it?" More to the point, who isn't? How well would your job survive a millennium meltdown?

To get the answer you have to start by looking at the most basic questions. We will come on to supply chain and office operations shortly, but even before you look at those, you have to know if you will physically be able to get to work.

The Smart Building

If you work in a modern building, then it is almost inevitable that some, if not all, of the building functions are micro-processor (i.e. chip) controlled.

Have you ever wondered why in some of the tallest buildings you don't have to wait very long for a lift, even when times are busy? The answer is that lift movements are controlled by chips that store information about how the building's visitors use them. These chips make sure the lifts are in the right places to serve demand, or travel express past floors when they reach a certain load.

The most advanced use of microchips in buildings is in employee smartcards. These are often used in tags that you wear around your neck, or passes that can be held up to pads on the wall or swiped through magnetic strip readers in the same way as a credit card in a shop. The most sophisticated ones don't even need to be swiped. They read who is wearing them and either automatically open doors, or bar entry. As most of these systems are fairly recent, you should find a high degree of millennium compliance, but as with everything in this book, take nothing at face value.

Don't forget, all the millennium compliant computers in the world will be useless if nobody can get to their desks to use them!

Everything in a smart building, from climate control and monitoring, to operating the fans and disinfectant dispensers in the toilets is run by computers and embedded processors. We take it for granted in high rise buildings that we will be able to work in a clean and comfortable shirt-sleeve environment, but that is only because there is a network of sensors and systems monitoring the air continually and switching the flow to ensure steady ambient conditions. If those start to go haywire, you could find your state-of-the-art, glass office becomes a high-rise greenhouse.

You need to ask your employers about the compliance of fire and security systems. Once you get in to the building, will it be safe? It is a very big job to find out, but it is your responsibility to do so, because if you don't make sure the questions are asked, and the right answers found, you could be putting *your* life in danger.

Let me explain why.

A company in the United Kingdom with a wide network of buildings decided that as most of its premises used the same building management systems, it would carry out millennium compliance testing on these systems in a building that had just been replaced by a new one in the same area.

Given that some buildings in use today can have thousands of chips in use for all their various functions, that was one almighty task. Fortunately they started early.

They checked out the alarm system and found it was fully

compliant, as were the fire sensors. However, so the story goes, one night during the testing programme, the smoke alarm triggered the fire system. All appeared to be working fine, until they came in and discovered that instead of isolating the building, the system had opened the gas main full on. Luckily the building had been isolated from the gas supply in the street, otherwise it would have made headline news.

The employees of that company should be alright, they happen to be a very millennium-conscious firm, but what about the people that own *your* building? How much have they done to ensure it will function properly? Have they issued you with a statement detailing their processes and controls for millennium compliance? If not, why not?

Once you get to work, what are you going to find?

Will your telephones work? Will your computers work? Will your factory equipment work?

Again the questions have to be asked by you. As the Xerox company likes to point out – 'You are your own safety manager,' and in the same way you are ultimately responsible for making sure you can do your work and earn your salary.

There are very few companies in the world that actually have large pots of money sitting in the bank on which they can draw to pay their staff. If they don't earn income, then they sure as heck are not going to have the money to pay either their creditors, or if the worst comes to the worst, their staff.

This Means You!

Every job in the world depends on a whole bunch of other people and companies. Your s does, and mine most certainly does —If I don't have a working computer, then I can't do research on the Internet, or send people e-mails with questions and get their replies. I can't even send and receive faxes because my trusty Apple iMac does all those things for me.

Luckily I don't need to worry about millennium bugs because the Mac is 100% compliant, but I do have to worry about my Internet link, and the telephone company that connects me with the outside world and transmits the signals from me to the 'web.

In practical terms if I received a commission from a magazine and failed to deliver, they would not care whether it was a technology problem, or simply laziness on my part, they would find someone else to write the article and that would be my last commission from them.

All businesses think the same way – reliability matters. So if you lose your reliability, don't be surprised if your business follows it rapidly.

Just-in-Time

No not the song, the business acronym – 'JIT.'

JIT stands for 'Just-in-Time,' which is the total antithesis of being prepared. It is a method of distributing things to arrive 'Just-in-Time,' for the moment they are actually needed, rather than holding stock to cover peaks and troughs. It is now used almost universally in every industry from food retailing to manufacturing.

For instance, if you go to a big car manufacturing plant , you will see that they actually make very little. Body shells arrive ready assembled and painted. Engines come in fully built-up, tested and even filled with oil, transmissions arrive from a transmission factory, seats from an upholsterer. The car manufacturing 'factory' is in reality nothing more than an assembly plant.

Holding all those components in stock would need

another two or three plants the same size, so manufacturers hold a tiny stock, sometimes only enough to keep the production line running for a day. Each individual supplier has timed delivery slots when their trucks arrive at the factories with enough to keep the factory supplied until the next delivery.

When it works, it is an extremely efficient system. Wastage is reduced, and the costs of maintaining large stores are removed. The end result is a cheaper product for the consumer, but it relies on everybody in the chain being able to react to demand at the end of the line.

It is of course entirely computer controlled, but when one part of the network slips up, the whole thing grinds to a halt. A typical case arose in a car plant last year where the wrong engines were delivered. The line had to be stopped while somebody went in, got the engines out of the factory and replaced them with the right ones. The whole process stopped production for three days, and affected the car manufacturer and all its suppliers.

Is your company part of a JIT chain? If so, can you guarantee continuity of supply, not just from your immediate suppliers, but from theirs? Certainly, as the millennium date approaches, more and more big organisations are becoming uneasy about their suppliers. One survey revealed that of the 15,000 companies that do business with one major global organisation, less than one-fifth had completed millennium compliance programmes by the turn of the year, and over a third had not even considered doing so.

So what can you do about it?

The best way to protect yourself is to make sure that you leave as little as possible to chance. If you are dependent on somebody else for your ability to work, then make sure that you won't be stopped from working by them fouling up. It's a simple rule, but one whose implications are enormous. You also need to get the people on whom you depend to do the same for the people up line from them and so on.

Here's a checklist, in no particular order, of some of the things you might rely on and ought to check out:

Item	Who to ask about compliance
Raw materials used by you	Your supplier, and his, and his.
Machine tools	Manufacturer
Distribution/Storage systems	Manufacturer/Computer system supplier
Conveyor belts	Manufacturer
Fork lift trucks	Manufacturer
Electricity distribution system	Building owner
Central heating/air conditioning	Building owner
Coffee machine	Manufacturer/Rental Company
Toilets (Fans hand dryers etc.)	Washroom Maintenance Company.
Computer systems	Your supplier
Lift	Building owner/Lift manufacturer
Security System	Building owner/Alarm manufacturer
Fire Alarm	Building owner/Alarm manufacturer
Telephone switchboard	BT/Switchboard manufacturer

If you work for a large organisation, you should be fairly confident that you have a millennium compliance programme manager. If you do, it would be worth asking him or her about your company's state of readiness. If you don't, now would be a very good time to start worrying!

You never know, if your company does not have a

millennium programme in place, you could take this book to your chief executive and find yourself with a new job creating one. That would probably be the most poisoned chalice you could ever sip!

Even if your company does have a millennium compliance programme, you need to think about how your company could get around the problems that losing some, any, or even all, of its computer and chip controlled systems would create.

It's an interesting exercise. Sadly a significant number of companies whose managers went through that exercise when the bug first started making the headlines, looked at the costs of compliance and realised that they just could not afford the investment. Their conclusion was that it would be cheaper for them just to go out of business.

Let's just hope you don't work for one of those companies.

If you do, now would be a very good time to start saving your pennies, stocking up and reading the job ads in your sector.

Sadly it's the lack of preparedness by so many organisations that is creating the risk of turning what should have been a wholly foreseeable and solvable problem, into an event that holds the potential for the a major loss of confidence in the way Society is structured today.

Read on to find out why.

7
How safe is your money?

A very good question indeed. If by now you have not realised that one of the most potent issues around the millennium bug is the safety of the financial services systems, then you have not been adding two and two as we proceed!

The financial services industry is undoubtedly the single most computer-dependent business sector in the world. It is also at the heart of the global economy. All money transactions are handled electronically. Interest charges and payments are calculated by computers. The whole system that starts by you paying in a cheque and translates into you being able to withdraw cash, is computer controlled. It all relies on confidence.

Financial Services is a very much bigger business than just banking. Not only is the entire stock market trading system in many countries now completely paperless, large chunks of insurance business are now transacted on-line, and almost all the records that the companies consult when they talk to you on the telephone are held and managed by computer.

When a telesales person says "I'll just get out your file," what she actually means is "I'll call up your record on screen." They stick with the old saying because it sounds more human.

Computerised records now cover every aspect of finance, from salary and tax payments, National Insurance contributions, VAT records, travel insurance, medical cover, home and contents insurance, to motor policies, commercial insurance, bank accounts, deposit accounts, mortgages, ISAs, PEPs Pensions, stocks and shares. Whatever you think of to do with money is now saved, managed or transacted by computers.

There are also an awful lot of chips in finance – Cash machines, security systems, safe door controls, video monitoring and security systems. Any of these could break down in the millennium and cause a major hold up to normal operations.

So if it's that important, they should have got it right?

If only…

According to all the computer experts, as we rolled into 1999, with the countdown ticking ever-more remorselessly, we should have achieved the point where everybody's millennium compliance testing programmes were finished and in place with plenty of time left to test them before we hit either the September problem or Millennium Day itself.

Let me reproduce the statement s made to Action 2000 in January 1999 by two of the Financial Services Industry's key players.

Michael Foot, Managing Director of Financial Supervision at the FSA:

"No regulator can guarantee business continuity. Year 2000 is a major supervisory priority for the FSA. Exchanges and clearing houses are broadly on track with their programmes. With regard to individual firms, our particular focus is on those with a potential

high impact on consumers and markets. The large majority of these are either on track for year 2000 compliance or, if behind, well-placed to catch up in time A few firms are in danger of not being compliant in time and intense supervisory action is being taken with these.

Author's comment, please note the use of the words: *'broadly,' 'large majority' 'a few'* and *'intense supervisory action.'* Please also note the total lack of definite numbers and percentages.

Alistair Clark, Executive Director, Financial Stability, at the Bank of England:

"Year 2000 is a big issue for the financial sector, given its heavy dependence on IT, and the interdependencies both within and outside the sector. The main infrastructure providers have been "on the case" for two or three years, sometimes longer; they have already undertaken extensive testing and provided extensive information about the actions they have taken, so that preparation in this area are well on schedule; but there remains more to do on testing, on risk mitigation and contingency planning and on the international front".

By the time this statement was made, the main infrastructure providers had been 'on the case' for two or three years. Yet they hadn't got to the point that the industry could turn around and say "We've cracked it." Does this mean that the companies who only started a couple of years ago are going to be so far back that they haven't a hope of catching up?

In fact when you read more of the statements you would find references to some works not being completed until June 1999 as final sign-off dates. I may not be a computer expert, but six months to the deadline really seems to be a very short time, especially when the computer people say that 80% of the time taken on any project covers the last 20% of the work.

Show Me The Money!

Unfortunately, none of these statements gives what any responsible journalist wants – a definite quantifiable statement of how well the sector is progressing and what it is going to do if it doesn't make the target. As Tom Cruise was told so memorably in 'Jerry Maguire' "All the talk in the world don't mean nothing 'til you got the dollars in your hand. Don't give me words – Show Me The Money!"

Although the public utterances of the banking sector seem very confident and self assured, when you dissect them, you find a lot of what the marketing industry calls 'motherhood statements' – soothing words that make you feel better, but in reality there's no substance. Just like mummy saying 'there there.'

They sure ain't showing nobody no money!

Underneath the calm surface, the financial services industry is doing an extremely passable impression of a duck paddling away like goodness-knows-what in a desperate effort to make sure it gets there on time. Certainly if you want to earn big bucks at the moment, the place to do it is in the City of London as a millennium compliance consultant.

As a last resort, there is the reasssurance that your savings and assets should not be lost as a result of computer problems. Banks and other financial service providers are extremely careful about ensuring that all their data is well and truly backed up, so that even if the systems do go out of action for a while, they will be able to restore the data when they come back on-line. This means that, much as you might like them to, the chances of the banks losing details of your loans and overdraft are virtually nil!

Although individually the financial services companies are doing their damndest to make sure the banking system doesn't come crashing down on 31December 1999, it is vital to remember that financial services is now a global industry in which everybody depends on everybody else. So, even if the UK achieves the target, if France or Germany or any of the UK's major trading partners doesn't get there, we could still face a major financial headache.

Remember also that there is a very large percentage of lending to foreign countries, which could be seriously prejudiced by their inability to beat the bug. If the economy of a country goes into free fall as a result of the bug, and one of our major banks is owed a huge chunk of money, then guess who could follow the said country in a downward spiral.

As an example look at the state of Honduras. Already struggling under global debts running into billions of dollar, the country's entire infrastructure was wiped out by Hurricane Mitch. Honduras today does not have any money, and the total loss of its banana crop means that it has no way of earning any. Its prime minister has already told the world banking community that it needs both the existing loans written off, and new loans to rebuild the country. But who's going to pay the banks?

If you don't believe that the banking industry could face a world-wide problem, then just look at Japan where 1998 saw the collapse of financial institutions that everyone had thought were rock solid, purely because some of the people to whom they had loaned money were no longer worth what they owed. In that respect, a country can be just as vulnerable as a company, especially if it has not prepared for the effects of the Millennium Bug. There are many countries where millennium preparation is simply not on the agenda, yet the effects of the bug could bring their economies to a total standstill.

So what are people going to do?

Knowing the human race, people are going to do two things. The first is try and ignore the problem in the hope that it will go away. Then, when they realise that the problem is going to stay put and in all probability become even worse, they will panic.

Predictions in the City range from a crash in the stock market, as people try and cash in their assets and hold them in something more tangible, to a run on the banks as customers demand cash. Even the most conservative millennium watchers are suggesting that you should take out extra cash to cover the first couple of weeks in January, in case there are problems with cash machines, or credit card readers.

Great idea, but remember that at any time less than one percent of the currency that is held in banks is actually available as printed notes. The rest is all 'virtual.' It has been loaned out to the bank's borrowers. The reality of your bank statement is that it is nothing more than an IOU notice from the bank.

Now, do a quick sum. How much cash do you normally take out of the bank in a month? Compare that to how much you spend on credit or debit cards and directly through the bank on direct debits or standing orders. For a lot of people the cash they withdraw is less than a tenth of what they spend through the electronic wizardry of the banking system.

Imagine if as Christmas approached, not just you but everyone else as well, were to try and take out enough cash to pay your regular monthly bills and tide you over in case you could not use your cards. The demand for cash in December would be so great that the banks could not cope. There could be some very nasty scenes as customers demand their money, but find that the bank did not have any to give them. The banks would quite literally be bankrupt.

According to one report, the Australian Federal Reserve has decided that it will not scrap any returned notes for 1999 but stockpile them instead. It is also printing a lot of extra money in anticipation in the hope that there will be no shortage of cash and no run on the banks. Another says the US Federal Reserve is printing an extra fifty billion dollars in notes in the hope that this will suffice.

In Britain the banks are politely telling us that there is no need to panic and that everything is under control, so we won't need to have any fears about not being able to withdraw cash in December.

I will however, certainly make sure that on the 31st of December I get my bank to give me a printed statement of how much they are holding in my account, and get the Bank Manager to sign it.

That's assuming the bank will actually open its doors, and I can get through the throng of people demanding cash. Mind you, there's another worrying thought. If a lot of people do withdraw cash, where are they going to put it to keep it safe. An awful lot of burglars might see the end of 1999 as their most profitable season for a millennium.

As long as they keep their hands off my food supplies.

"What supplies?" I hear you ask.

That's the subject of the next section.

8
Food and Drink

Man has three basic needs: Food, drink and warmth.

At a pinch most people can find warmth, even in a European or North American winter, through staying indoors and wrapping up. The old and the very young are most at risk from cold weather, so if things do go awry, they will need to be looked after.

Food and drink are more difficult. We talked about water in an earlier chapter, and saw that although the monitoring and management of water supplies is highly automated, the mechanical process is fairly straightforward, and unless you are hit by drought you should be able to collect enough to live on.

Food unfortunately is not so readily available. Although if is comforting to think of fields full of crops, being sown and harvested by ruddy-cheeked farmers who can spare a few sheafs of corn, the reality is very different. Farming at the end of the twentieth century is not so much agri-culture as agri-business. Today's farms are just as hi-tech as the rest of industry.

the dangers of food scares, whether from ingredients, GM products or tampering, mean that supermarkets demand their suppliers keep a detailed record of the whole life of every plant in their field. These include measurements of the amount of fertilisers and pesticide that each one has received. They can track which batch of seed was used and the supplier's name. Of course all this information is computerised. Farming is now a very computer-dependent business. Crop planting patterns are planned by computer, the amounts of fertiliser to be used are determined by computer models, and the crops are managed to ensure they ripen at just the right time.

If you go into your supermarket at any time of year you will find fresh fruit and vegetables regardless of the season. Look to see where they come from and you will find just about every exotic country in the world. Most fruit and vegetables are shipped across the oceans, many of the more luxurious items are air freighted. If those supply lines dry up, the produce shelves in the supermarket will be very, very empty.

Although produce is eagerly presented as fresh, and indeed is often delivered to the store less than 24 hours after being picked, food as a whole is nothing like as 'natural' as the people who produce it would like us to believe. Their image of home cooking using freshly harvested, natural ingredients to provide you with the most tasty, wholesome and enjoyable dishes right to your table, is precisely that – a carefully nurtured image.

The reality is that the food industry is one of the most automated and process-driven businesses you could ever find. I once took a trip around a food processing factory where the boss showed me with delight how the ingredients went in at one end, were cleaned, processed and combined along the stainless steel tubes that also served as cookers, until oozing out the other end where they were dispensed into metal containers, sealed and boxed, then dispatched to the food stores.

Suffice to say, that since then I have avoided eating ready-prepared meals, and I view somewhat cynically claims like '100% beef.' The products may indeed contain 100% cow, but not necessarily parts of it that I would like to see sitting on my table to be carved for Sunday lunch.

The food industry is a giant automated machine driven by the need to provide the consumer with a consistent quality of product that looks and tastes the same. It relies on technology at every stage to ensure that the exactly precise measures of ingredients are used, from flour to emulsifiers, cheese powders to stabilisers. It is not so much a home kitchen on steroids as a giant chemistry set.

The food chain of today is incredibly vulnerable to disruption, and one of the most potent examples of Just-In-Time logistics you will come across.

Over 90% of food in the United Kingdom is sold through the supermarket chains. They have been able to achieve this high level of market presence through a slick combination of monitoring purchase patterns, presenting a clean and well-managed shopping environment, and offering good value for money.

Behind the scenes is an exceptionally sophisticated logistical leviathan that runs something like this: When you go into the supermarket and do your shopping, all your purchases are scanned at the checkout, you are presented with a bill, you hand over your loyalty card, pay your money and go on your way. As far as you're concerned, the deed is done and you can rest easy for another week.

From the supermarket's point of view, the work is only just starting. The details of everything you purchased were recorded by the barcode scanner at the till as a data file. Every so often, the till uploads the data it has collected to another computer, called a server, which distributes the data to various systems used by the supermarket group.

One of these is the central buying function. This collates the sales data and uses it to forecast demand for the next day. Every day, sometimes every few hours, the central buying department computers generate orders that are transmitted electronically to the supermarket's suppliers.

The supplier orders are received electronically and drive

their production process. In some cases there is little human intervention, it is so highly automated.

Once the foods are manufactured, often within hours of the orders being transmitted, they are moved by road to the supermarket's distribution depots, where the orders are broken down into individual consignments for each store, then despatched again to the shop.

Notice the absence of one element in that chain? – Stock. That's right. There isn't any.

90% of the United Kingdom's food stuffs are bought, re-ordered, manufactured or processed, and re-stocked on supermarket shelves within a 24 to 48 hour cycle.

That is true for most things in the store from fresh eggs to baked beans, coffee, tea, beer, wine and spirits, dough products for in-store bakeries, drug companies for in-store pharmacies, dairy products, fruit importers, right down to cabbages picked the day before from farmers' fields.

You may have noticed that sometimes your local store tries out a new product, then you go back a few weeks later to find they no longer carry it. You may well ask the manager about it, and receive his assurance that they will try and get the product for you, but the decision whether or not to devote shelf space to a particular line is not his to take. It is ultimately made by a computer that analyses the volume of sales of every product the store carries, and decides what can be carried on the shelves most profitably.

The implications for you...

...are pretty simple – If anything goes wrong in the order/ manufacture/ distribution/ re-stocking chain, then you will find your local supermarket noticeably lacking in food.

It may not affect the whole contents of the store, but there's nothing quite so likely to send people into a frenzy of panic buying as seeing empty shelves in supermarkets. If you've ever tried shopping in one on Christmas Eve, you'll know exactly what I mean.

Over the millennium period, many of the major groups are planning to shut the shops for as long as three days. Given that there will be all the usual pre-Christmas stocking up going on anyway, and that people will suddenly wake up to the possibility of millennium bug related problems, I would suspect that December in the supermarkets will be pandemonium.

I bet you won't be in the least bit surprised to hear that the supermarkets' publicly expressed view is that it will be alright on the night. One major group is advising its customers that there really is no need to panic. In fairness their millennium preparation is very advanced and they have taken the matter very seriously, but even they are only too well aware how easily they could be let down by their suppliers.

There is actually not very much that supermarkets can do to protect themselves or their customers. Their business is totally driven by turnover, and they don't have the facility to hold supplies in stock, so lack of deliveries will mean empty shelves.

They are however working behind the scenes to avoid panic buying. You cannot but have noticed how much effort they have put into getting their customers to sign up for reward cards. You may even have one yourself. If you did not think too much about it, you probably thought this was a good idea to encourage your loyalty by giving you back a small percentage of what you spend.

To the supermarkets, the cost of giving you back a few pennies in the pound is peanuts compared to what you give them by signing up to the scheme. To start with, your name goes onto a database of customers, and a file is built up that shows everything you purchase in the store.

Initially the stores use this to track purchasing habits and identify the types of products that particular types of people buy. It helps them to forecast with some degree of accuracy how well a particular brand of food will sell in different locations.

It will ultimately become a massive marketing tool to sell you products, perhaps by sending you discount vouchers. For example, let's say you started to buy disposable nappies. The loyalty card computer would notice this as an unusual spending pattern and watch to see if you do it again. If you do, it would check on the size , and if they were 'new-born,' it would create a note on your file that either you or somebody close to you had just had a baby. You would never know, but coincidentally you might receive a booklet of discount coupons from the store containing offers on a lot of baby related stuff on which you could get a handsome discount.

If it was your baby, you might get a warm fuzzy feeling about the supermarket. If it was somebody else's you would pass over the coupons and the supermarket would gain another customer.

Nothing wrong in that you think. But what if you were to receive a letter from the supermarket out of the blue saying that they have been looking at millennium bug problems and there really is no need to panic?

You might not think back to realise that you had been putting a few extra cans into your regular shopping, but that's what the computer would be reacting to. In fact, many supermarkets are already monitoring customer card purchases and have plans to prevent people hoarding.

It does not take much to imagine how things could turn out later in the year as the millennium day draws near and people find themselves being rationed in the supermarkets, even before the bug strikes.

So how do you get around it?

Whether as a result of the Millennium Bug, or simply because of pre-millennium panic buying engendered by press stories and supermarket tactics, there are going to be disruptions to the food supply chain.

At the beginning of 1999 the chairman of Action 2000 was widely reported as having said that people should have at least two weeks worth of food and medicines in stock. Reports on the Internet attributed a Red Cross spokesman as suggesting a month, although nobody has authenticated either of those quotes. There are plenty of people in the United States who are so concerned about possible disruption that they have laid in a year or two's supply.

Whatever you decide to do, it seems obvious that building up some food stocks would be sensible. The problem is deciding what you should get in. Certainly you want to buy things you can preserve for a long time, and enjoy eating, because if the problem turns out to be a lot less than you expect, you don't want to be left with a load of stuff you'd never eat anyway!

As to how much you buy – that has to be your own decision. There are schools of thought that say you should have at least a month's supply. Some go to a whole year, while others say that if it gets to the point that you have to live off stored food for a year, then feeding will be the least of our problems. As I say, it has to be your own decision.

You also need to think carefully about the things that you eat normally and try not to stray too far. There is a world of difference between the romantic notion of home baked bread, cakes and pies fashioned by mother's fair hand over her log-burning stove, and the convenience of popping into the shop for a ready-to-cook meal. You need to buy things that are not going to call for you to make a total lifestyle change. If you're not a Delia Smith or a Mary Berry, then don't expect to turn into one overnight, especially if you have to do so as a result of the power or gas going down!

When you have decided what to buy, don't rush out and buy everything at once. Apart from the fact that it will draw attention to yourself, either from other shoppers, or from the supermarkets themselves, there is an obvious capital cost that you will incur. Instead, decide what you are going to store and acquire it over the course of a period of time.

One consideration with food is sell-by / best before dates. Even if you can get your hands on an abundant supply of fresh vegetables, they will only last for a week or so at best. You also have to be mindful of the possibility of power disruption – the electricity would only need to be off for a day, and the air temperature somewhere comfortably above freezing, for all the food in your fridge and freezer to go off. Even on tins, you need to check the dates and buy sensibly.

If you have alternative cooking facilities don't over-stretch them. Forget lavish dinner parties – If the gas is off, you would be better advised concentrating on simple meals.

If you don't have an alternative cooker, there is little point in depending on things that you have to cook. Having said that, you can cook food in a can, provided you pierce the top to let air out, but it can leave a nasty taste in the mouth, and you might have a problem telling when the food is fully cooked.

You also need to be aware of nutritional values, and the importance of maintaining vitamin intake. Just because your stomach is a bit less full than you might like does not mean you are in danger of starvation! The body can store up to a month's supply of most vitamins, but first to go is Vitamin B. Consult your pharmacist about a good source of multivitamins, but again, check the sell-by dates.

Whatever you choose to buy and store, you are better to go for straightforward products rather than heavily processed foods, because you will be able to use them in recipes as well as eat them in their ordinary state. Corned beef is an excellent example—you can make an excellent corned beef casserole or use it in sandwiches or fry it, whereas a tinned beef pie is precisely that!

If you decide to go for a very self-sufficient approach you might look at storing things in grain form such as Coffee beans or wheat grains, rather than the powdered variety. Grains, pulses and beans will last a very long time if you keep them absolutely dry. Make sure you have a grinder that doesn't rely on electricity, and that you buy enough to supply your needs. Remember when you grind corn or coffee beans down, the volume of flour or powder is a fraction of what you started with.

The checklist of foodstuffs at the end of this chapter might come in useful as a guide to the sort of quantities you might need for a period of disruption. Feel free to add to it or throw ideas out. It's by no means exhaustive, but it should help to get you thinking.

Storing food safely

Once you start to lay in a stock of food, you have to make sure that it is only used to feed you and your family, and not become a haven for creepie-crawlies, or rodents. If you have children, I would also advise you not to bring to their attention that you are laying in stores. Children are totally incapable of keeping secrets, and there are some things you don't want broadcast.

Always store food in cool dark dry places, and keep it off the ground. Moisture and heat cause bacteria and lead to the growth of mould. Make sure that all the containers you use are both insect and rodent roof. Tins and bottles are obviously a lot better than plastic, but make sure they are thoroughly sealed. Use adhesive tape to get a good seal and if you take something out of a container, always re-seal it with tape. Not only will that stop the air getting to your supplies, it will also reduce any smells and stop rodents or insects from knowing that they are edible.

Stock rotation and orderly storage are also very important. You should turn cans over every two or three weeks so that the contents do not settle or separate, and you should store like products together. That way if you lose a label, you still have reasonable chance of identifying what is inside. Another good idea is to mark the tin lid and bottom with a felt-tip marker.

One of the great things about tins is that they normally have a shelf life far longer than is shown on the label. Food manufacturers use different methods for coding cans, but the general principle to follow is that if the can is intact, it will probably be good for as long as five or six years after the best before date. However, use your common sense. If you come to use it and find the can has been punctured, is rusty or smelly, or the contents appear to be in anything less than perfect condition, THROW IT AWAY – you do not want to risk food poisoning.

The cooler the location, the better stores will keep. Cellars and lofts both have advantages, although cellars can be damp. Lofts have the benefit of being dark, unless there is a skylight. If there is, then blank it out. Another consideration is that although a loft has the advantage of letting you keep your stores discreetly hidden, it does mean that everything has to be taken up and down by ladder.

It is also important if you are going to use your loft, to make sure that you have adequate floor boarding over the rafters – the last thing you want is to find your stores coming crashing into your bedroom, or even worse you put your foot through the ceiling while trying to find a tin of baked beans!

Be very careful about what you store and how you do it. Remember that rodents are partial to a lot of different things – they will even eat soap. The last thing you want to share your food with is the local population of small furry creatures.

When you're out shopping for stores, don't forget the family pets. If things do go a bit pear-shaped, you might well find that your trusty pooch is the only alarm system you have, and if you start to run out of food you won't have many scraps left for him.

Why is an alarm system important? Well that depends on your view of law and order.

There are some interesting perspectives on that subject in the next chapter after the suggested food stores

Six months' food supply for two people

When you go out and buy everything you need for a family and put it all in one place, you will be amazed not just at how much it costs, but also at how much room it takes up.

If you think the problem is only likely to last for one month, divide the quantities by 6. Halve them if you think it will be a three-month problem, and so on to suit yourself.

Don't forget to rotate stocks so that you use the stuff with the closest 'best before' dates.

If you have a family of 4, double the quantities.

The checklist makes a few basic assumptions. Not least that you will be able to grind corn and do your own baking. It's the sort of shopping list an American trapper might put together to go and spend winter in the wilderness, so tends to be high on basics and short on luxuries.

If you want to take self-sufficiency seriously, then I would advise you to make use of the summer to get used to doing this while you have all the facilities of modern living still easily to hand.

Don't forget to try and vary the menus. There are quite a few staples on this list that should let you enjoy a good degree of variety. However, you will have to be careful and ration the foods you use. Especially if the situation continues for along time.

If things get really bad, then hopefully by the time you run out of this lot, you should be growing your own vegetables, and you will have found ways of getting meat.

Don't forget to think about your pets and get in plenty of food for them as well

Grains
70 kg pounds of hard wheat or a combination of 35 kg wheat and 35 kg of flour.
15 kg dry corn to grind for cornmeal
15 kg soft wheat
15 kg white rice
15 kg brown rice
15 kg oatmeal
15 kg corn flour

Pulses
15 kg of beans, such as broad beans, kidney, etc.
5 kg of split peas
5 kg lentils

Dairy
Large block of Cheddar Cheese (will keep well if cool and well wrapped)
12 cans or boxes powdered milk
2 large cans cheese powder
5 large cans dehydrated eggs
3 large cans butter or margarine

Sugar
12 kg white granulated sugar
2kg brown sugar

Shortening/Oil

12 litres vegetable oil
2 litres olive oil

Salt

2.5 kg iodised table salt

Fruits

26 tins peaches
26 tins fruit cocktail
26 tins apples 26 tins pears
52 tins miscellaneous. fruits
1 large can raisins
1 Large can dehydrated strawberries
6 Large cans dehydrated apple slices (use these for making pies and sauces as well as to eat)
2 Large cans dehydrated banana slices

Vegetables

40 tins of green beans
40 tins of sweet corn
40 tins of carrots
40 tins of tomatoes
40 tins mushrooms
20 tins Red kidney beans
20 tins tomatoes
30 tins baked beans
40 tins of other vegetables you like
6 litres of tomato sauce
12 x 2.5 litre tins potatoes
5 kg instant potatoes
1 Large can dehydrated onions
2 Large cans dehydrated broccoli

Pasta

10kg spaghetti
6 kg assorted noodles
6 kg lasagne

Meat

36 tins corned beef
36 tins chicken/turkey
36 cans ham
36 tins tuna
36 tins Spam
20 tins red salmon

Seeds

A large selection of garden seeds to replenish your food
supply, should the period of hard times last longer than a
few months. Always opt for the worst and prepare ahead.

Most garden seeds last for years, if kept dry. One notable
exception is onion seed, which should be replaced yearly.

Miscellaneous

0.5 kg baking soda
1 kg baking powder
0.5 kg dry yeast
A selection of spices that you usually use
6 Large jars coffee
1,500 tea bags
A grain mill to grind grains
A cooking with basics cookbook
1 gallon syrup or treacle
An assortment of "treats", such as pickles, jams, preserves
Vitamin tablets
Chocolate
Assorted crisps
Breakfast cereals
Packet puddings
Tinned sponge puddings
Tinned creamed Rice / Semolina

9

Law, Order and survival

This is the part of the book where it starts to get scary. If you've followed it this far you will probably be reading on with some trepidation. I'm sorry to tell you that if you are concerned, you are rightly so.

The whole of the emergency services infrastructure depends on computer technology. Police cars, fire engines and ambulances are all sent to incidents by radio, whose signals are generated in highly computerised control rooms. Their locations are often recorded on the controllers' screens so that resources can be mobilised to deal with incidents as they arise. You will even find that some health authorities now station ambulances in various locations around towns and cities rather than in a central depot, so that they can respond more quickly.

As technology has improved, its uses in the maintenance of law and order has increased dramatically. Town centres are now no longer monitored by patrolling policemen, but by CCTV cameras whose signals are sent back to the police control rooms. Motorways use the same technology, with the added use of speed cameras to deter motorists from going too fast. Criminal records are held on a national database that can be accessed by police officers all over the country to track the movements of known criminals and solve crimes more quickly.

If the computer system goes down, then so do the eyes, ears and intelligence of the emergency services. Their ability to respond to calls and solve crimes will follow immediately. An example of that arose in New York City at the end of January when a routine test cut off power at the 911 call centre and a back-up system failed to start. Apparently the power was shut off at the city's main 911 call centre as part of a test of its ability to withstand power cuts. However, the emergency generators did not work, so the call centre went off-line. For a period of about an hour the whole of New York City was without emergency services cover. Although this glitch was not attributed to the millennium bug, it shows the vulnerability of the system.

Fortunately for New York, the glitch occurred on a quiet part of a Sunday morning. Even then it was reckoned that about 500 emergency calls went unanswered.

The fact that the United Kingdom's police force has had all leave and rest days cancelled for the period leading up to and over the millennium night, as has happened in many US States and European countries, coupled to reports of the National Guard being put on standby, and the mobilisation of Canada's forces, suggest that the authorities are taking the threat of disruption extremely seriously. Given the vulnerability of their infrastructure, they would appear wise to do so.

Law and order relies upon the overpowering desire of the populace to live within its rules. Unfortunately as the twentieth century draws to a close, the fragility of those rules, and the fact that their observation is primarily voluntary, is made ever more obvious by the growth in violence and street crime, the loss of respect for authority as a whole, and the sickening rise in attacks against the elderly and less able members of society.

Are we prepared?

In a word. No. As the deadline draws near, the emergency services are dusting off their major incident plans and looking to see how these would be able to handle the severe disruption of society following a temporary, or even sustained collapse of the infrastructure.

In the dark days of the cold war, Civil Defence groups were formed, and a substantial set of emergency plans were developed to maintain law and order in the event of a nuclear attack. To a great extent, these plans relied on the nation's citizens playing their part and living in an orderly and disciplined manner. To make sure that they did, soldiers were to be stationed at key points to control crowds and prevent looting, the motorway network would be closed off and used either as strategic military airfields or as fast routes for military traffic, and the populace would be encouraged to stay at home and wait for news.

You can almost hear the Second World War jingoism. In fact these plans were current until about ten years ago. No doubt somebody in government has already dug them out. Or should that be 'hopefully' somebody in government has dug them out, because there is very little suggestion coming forth from the public spokesman in government circles that they are even contemplating the problem.

Looking around most local authorities, there is a frightening lack of preparedness. The message from government that everything will be alright seems to have been accepted by them all too readily. Press reports that Town Halls are to be 'named and shamed' seem to have had little effect. In reality very few councils have either the budget, the know-how, or the spare manpower to do much other than try and bring their internal computer systems up to some sort of state of readiness.

Certainly the major institutions that could be disrupted are preparing emergency plans. Hospitals, as we have already seen, are getting in supplies and setting themselves up to be capable of standing alone, but as one Health Service Manager told me. "If the Bug really hits the power or gas we're going to be up to our necks in it. We'll be the only building for miles with lights on, so we'll be warm and we'll have food, but we'll stand out like a beacon in the darkness. We're going to have the devil's own job keeping people out."

Apparently that particular hospital has already had several meetings with both police and army officers, and is well on the way to preparing for what its manager called "the worst siege in this area since Bonnie Prince Charlie arrived."

In America there have already been ominous statements from prominent politicians and government managers. One senior member reported in January that the government would be moving from contingency planning to a crisis-management phase. Responding to a question about electrical power failures, he said, "In a crisis and emergency situation, the free market may not be the best way to distribute resources. If there's a point in time where we have to take resources and make a judgement on an emergency basis, we will be prepared to do that."

As you can guess the doomsayers in the US are seeing that as a public statement that if it all goes to hell in a hand basket the government will step in and declare martial law. Certainly the old Civil Defence scenarios viewed that as a logical state of affairs in Britain. No doubt the same situation applies in the rest of the industrialised world.

So what will happen?

As this book went to press it was being reported that one in three citizens in the US, Canada and Australia were convinced that there would be a breakdown of law and order once the bug strikes.

Their view is that without the infrastructure of law and order to restrain them, and bear in mind that this could be totally disrupted by something as simple as the failure of a communications centre, the less orderly elements in society will set off on a spree of looting and rioting.

It does not take much to get some sections of society going, as the citizens of Brixton, St. Pauls, Paris, Atlanta, Los Angeles and Bonn can testify. Even the annual celebrations of the New Year in London, Edinburgh and New York have been noted for their ability for drink to overcome common sense. If the power fails and the lights go out, God alone knows what could happen. The police would have their hands full just trying to get the celebratory crowds moved safely, let alone dealing with petty crime. Certainly if that did happen, the emergency services would not be able to cope, and it would be open season on shop premises for anyone with a torch.

If the infrastructure comes back on-line quickly, then the problems could be contained, hopefully before too much damage is caused. However, if supply line disruptions result in ongoing food shortages and panic in the stores, then you can be pretty sure that the people who are least prepared will be those who are most used to getting their way by physical persuasion rather than restrained discussion.

If disruption continues for any significant length of time, then restraint will reduce even further as hunger takes over, and you will find that your carefully preserved food stocks are at risk. Hence my earlier comment about not telling the children, and keeping your supplies out of sight. Some commentators have gone so far as to suggest that if things get really nasty, we will see organised groups coming out of the cities to pick on the countryside and raid it for supplies.

It is also worth remembering that if there are major supply line disruptions, one part of society that will be just as affected as any other will be the prison population. We've not seen prison riots for some years, but if the inmates of the jails are not fed and looked after as they are used to, you can be certain that somebody, somewhere will have the enlightened idea of letting them out to fend for themselves.

It is an awful thought that just as mankind is entering the information age, with the chance that everyone could be empowered by technology, that very technology threatens to change the course of history. The prediction that it will be 'every man for himself' looks horribly as though it could come true.

Sadly the evidence that it is happening was already there to see at the beginning of 1999. Among the plethora of websites dedicated to millennium horror stories was a site listing sources of supply and giving similar sorts of ideas to this book. To access it you had to pay a fee and become a member. I didn't join, so I don't know what they offered, but after they reached a certain number they pulled up the drawbridge and would not let anyone else in.

What can you do about it?

Obviously the first thought is to protect your family, the next is to protect your community. In many cases the two objectives will be best served together. Many such groups have already been formed as people wake up to the reality that the Millennium Bug could be extremely bad news. Some are moving out to the country, others are simply gathering together to share their resources and buy supplies to tide them over.

Certainly there is strength in numbers. If you have an extended family, this would be a very good time for you all to sit down and discuss your views of what could happen, and what you feel as a family you should be doing about it together. If you've read this far, you should have a pretty well-formed idea of how badly you think society is going to be affected.

Whatever you do, be extremely careful who you tell about it. Throughout society, whether in the built up areas of the cities, or the leafy glades of rural villages, there are people whose view is that if you have something, then they should have it too, even if that means depriving you of it! They will have little inclination to talk, so you my find yourself having to choose between sharing your food voluntarily or being forced to.

One of your best defences is for them to think you're in the same boat as they are, but inevitably if you are running a generator, or there is a cloud of smoke coming from your chimney, you will attract attention to yourself, and it may not be welcome.

The security of your home is going to be a very important issue. If you decide to have people to stay with you for the millennium night, then make sure they are part of your plans for the start of the new year. You may well decide to have a smaller gathering of very close friends or family who are prepared to stick together if the lights go out.

Whatever happens, do not flaunt your stores. Keep them hidden, and if possible split them up into caches around your house and property. That way if you are forced to give up some of your supplies by a mob of thugs, you should still have some left that they did not find. If you lay low afterwards, then hopefully they will think you have been cleaned out.

Keep some stores in an obvious place and be prepared to sacrifice them. You might even have a second line of stores in a less obvious place that you are prepared to sacrifice. Leave a simple trail that the thugs can find and feel pleased with themselves for doing so. The important thing is to keep your main stores very well hidden and camouflaged so that they are not easily found. Remember if it comes to mob rule you won't be dealing with intellect, just hunger.

Surviving at home

Whether you are expecting the descent of society into anarchy, or just the possibility of power or gas being out of action for a while, there are a number of sensible precautions you should consider for your home. You can look upon it as either serious preparedness for the worst, or like packing everything up to go on an extended caravan holiday. The difference is you don't go anywhere!

Some people are taking to the hills. Certainly in the US and Canada, where there is a strong leisure culture of getting back to nature, there are many city dwellers who are gearing up for the whole of January, or longer, spent in a cabin in the backwoods. In Europe we don't have the same sort of culture, so most people will just stay home.

You need to think about how your family would survive without the basics you take for granted – power, gas, fresh water and a ready supply of food. Whether or not the infrastructure of society disintegrates fully, there is a high probability of disruption, so being prepared has a lot of advantages.

If it turns cold, and we are talking about this all happening in the middle of winter, you need to make sure your family is all equipped with warm sturdy clothing. Good hiking boots, jeans, jerseys, fleeces, body warmers, thick socks. They could all be very useful.

If you can, you should either open up a fireplace or get a multi-fuel stove. Logs are a good source of heat, but tend to give off a lot of smoke, whereas you can get coal that is smokeless. If the worst comes to the worst, you can burn just about anything in a multi-fuel stove.

You may need to change your domestic arrangements. Heating a whole house is impractical if there is a fuel or power shortage, so you may decide to all congregate in one room. Choose whichever room has the most warmth, and make sure that you insulate it fully. Although it is a good idea to tape up windows to eliminate draughts, make sure that you don't allow a build up of carbon monoxide from an open fire. It would kill you silently in your sleep. Better to shiver a little, take a hot water bottle to bed with you, and huddle down in your sleeping bag.

However tempting it may be to throw a mattress on the floor, don't. Invest in camp beds and thick sleeping bags. Not only will you be kept above the ground, thereby avoiding heat loss through the floor, you will also be less likely to suffer from damp being drawn up through the floor by the warmth of your body.

Fitness

As a nation we have all tended to get somewhat out of condition. If we are forced to return to a less mobile society, then you don't want to find that you are out of breath the first time you try and go anywhere without the car.

Living without the luxuries of life is hard work, and before you try it, you need to be fit . With a whole summer ahead of us there is plenty of time, even if all you do is take up walking regularly and climbing stairs instead of taking lifts. Whether or not you need your fitness in the New Year, it's probably worth doing anyway!

Dig out your old bicycle and get back into condition over the summer. All the 4x4s in the world will be useless without petrol, and it takes a very long time to get anywhere by walking. Motorcyclists will of course see the millennium as a perfect excuse for buying an off-road motorcycle, or an ATB. If we do end up with villages and roads being closed off, it could be handy to be able to ride across country.

Hygiene

One of the biggest problems you hit when there is a restricted water supply is preserving hygiene. We all use a great deal of water – every time you prepare food or use the toilet you wash your hands. If you do this under a running tap, you can waste half a gallon of water.

If you do not take hygiene very seriously, then you open yourself up to all sorts of potential hazards. Any problems with sewerage, contaminated water or the build up of rubbish provide ideal breeding grounds for germs, bacteria and rats.

If your domestic water supply is cut off, you will not be able to wash your hands with such frequency or flush toilets, so you will need to look at alternative arrangements. It is absolutely vital that you take care of hygiene, especially as there is a very strong chance that you will be coming into contact with many germs that would not normally enter your life. Hands can be washed in a basin of water treated with a disinfectant. The basin can be shared or re-used. Disinfectant wipes are one alternative, but are expensive.

Camping and caravanning shops carry an excellent range of chemical toilets. These can be used to handle the solid wastes of a family, which can be disposed of by digging a hole in the garden and burying them.

If you don't have a chemical toilet, but find yourself wishing you had, the other alternative is to dig a pit in the garden, and fill it in once it has been used. Time consuming I know, but it could be your best alternative.

Liquid wastes are normally sterile, but do tend to have an odour. These can be disposed off outdoors, and in this respect men have a built-in advantage. If you dilute them they can be used as a nitrogen-rich fertiliser. If possible, avoid using the chemical toilet for liquid waste – this will just fill it up quickly and add to your disposal problems.

Security

An Englishman's home may historically be considered to be his castle, but there are very few of today's desirable residences that come complete with drawbridge and moat. In fact most modern homes are only marginally more secure than a tent, and if somebody is determined to get in, then a brick through a window is as effective way of doing so as any other.

How well you are prepared to protect yourself is entirely down to you. Fortunately the law in many countries now prevents people owning hand guns, so we should not see a return to the days of the wild west, at least not in Europe. However, if push comes to shove, you need to decide how far you are prepared to go in defence of your family.

William Golding's 'Lord of the Flies' paints an excellent picture of what happens to a society when the rules break down. I'm not saying it will be that bad, but urging you to think about it and make up your own mind.

Remember if the millennium bug does strike at people's jobs, their income and their food supplies, there will be a very large number of very unhappy people in the country. We have lost many of the traditional skills of frugality and self sufficiency that were handed down to our parents and theirs before them. Social unrest seems to be pretty much a certainty if the Bug does even a fraction of what it might.

Obviously there is a lot to be said for pulling together, but if the rule of law and order goes totally out the window, then it will be every person for him or herself. How likely you think that is to happen is something that only you can decide.

How you might reach that decision is the subject of the next chapter.

10

How bad will it be?

There's an old saying 'for want of a shoe the war was lost.'
To anyone below a certain age that sounds seriously weird,
so let me explain.

In the days when wars were fought by armies of men
slugging it out hand to hand, they would be won or lost by
the cleverness of the general leading the army. The saying
is part of a poem that starts something like 'For want of a
shoe the horse went lame.' It then goes on at length to
detail the train of events that followed, but to save time, I'll
sum them up – The horse went lame because its shoe fell
off. It fell over injuring its rider who happened to be the
general. His officers administered first aid and bandaged
his broken bones so he eventually managed to make it to
the battle, but by the time he got there, it was all over bar
the shouting and his side was royally defeated.

The moral of the tale, which comes from centuries ago, is
that the smallest, and apparently most unrelated error can
cause a major mishap. It's Victorian equivalent was
something about a spanner in the works bringing the whole
lot grinding to a halt.

The problem with the Millennium Bug is that everybody
agrees it has the potential to be an extremely large spanner.
But nobody knows for certain where, or how badly, it will
strike and how this will change society.

The vast majority of the world's population is suffering from the Willnott syndrome. – They honestly believe "It will not happen to me!" In January 1999, Gwynneth Flower, who heads up the British Government's Action 2000 campaign was quoted widely as follows, "All our efforts have failed to scare or energise people into action. The dilemma is how to raise anxiety levels without creating public panic."

In fact there has been a mass of articles in newspapers and magazines, and by the time you read this the broadcast media will have cottoned on as well, but the vast majority of the population, and here I'm talking in the upper 90%, sees the whole thing as some sort of business computer problem with very little personal relevance to themselves.

Once people start to realise that behind all the hype, the Millennium Bug represents a real and very tangible threat to society, they go through a series of stages. Awareness leads to disbelief. Disbelief becomes fear of what might happen. Fear turns into action. However, if too many people leave it too late to act, the result will be panic.

How is the world preparing for the crisis?

The answer to this question has to be – 'Not very well.'

Probably the most technologically aware countries are the United States and Canada, where the major global businesses have recognised and taken hold of the problem, so have a reasonable chance of making it through unscathed. Sadly, there are many countries whose economies are either so backward or debt-laden that they simply do not have the resources to spend on the problem. One estimate is that well over 100 sovereign states will hit the Millennium deadline with little or no effective preparation to speak of.

You can understand why in America survivalism has never been so popular – Hundreds of thousands of citizens are waking up to the thought that their cosy middle class lifestyles could be very rudely disturbed in a few months time and looking longingly back to a more self-sufficient society.

Russia only started work on its millennium problems at the beginning of 1999 and faces a bill estimated in the region of US$ 3billion. Although it occupies a huge part of the world's land mass, economically Russia is not the powerhouse it has always yearned to be. It is a very poor country, as are many parts of the former Soviet Union, so priority tends to go to everyday things like feeding the population

Certainly there are great tracts of the third world that will not achieve millennium compliance, but they are often far less reliant on computer technology, so the day-to-day effect on these countries could be much less than in the industrialised nations.

Where the problems will be most acute is in the interface between countries, whether through telecommunications systems being unable to interconnect, or through problems with shipping goods between them.

The knock-on effects of the bug could conceivably cause many more problems than the bug itself. The global economy is highly dependent on an intricate chain of raw materials from second and third world countries, being processed either in second or first world countries, and then sold to all nations. Many of the world's largest companies have been investing billions of dollars in second and third world countries in the hope of reaping vast rewards as their economies grow. If the bug upsets that particular applecart, which seems highly likely, then there will be many companies facing substantial losses on overseas operations. Many will react by closing those operations down, kicking off the start of a very deep recession. Very few countries are capable of returning rapidly to a state of self-sufficiency, so the possibility of food shortages, liquidations and company failures leading to mass unemployment and civil unrest start to become all too likely.

Overall there are very few parts of the world that make headlines for being well ahead in their preparations not just for the bug, but for its potential fall-out. In fact the opposite is true. Almost without exception, the world is far from ready to face this totally immovable deadline.

The three scenarios

There are three possible outcomes of the Millennium Bug. Each one has an equal likelihood of happening. Let's start with the easy one.

BUSINESS AS USUAL

This is the 'You wish' scenario.

In it, all the companies in the world will have tracked down every critical bit of code and put them right. There might be a few computers that don't make it, but by and large we'll come through alright. Possibly there might be some power outages, but not too many. There might even be a few lines in the shops that run out, but It Really Won't Be A Problem.

And we'll all live happily ever after in Cloud Cuckoo Land.

As you can probably guess, I don't quite see it that way, nor do many of the informed people I've spoken to, so let's look at the one that has got the doomsayers all excited.

TEOTWAWKI

As you know the computer industry is incapable of spelling anything out, if it can come up with a snappy acronym.

TEOTWAWKI stands for 'The End Of The World As We Know It,' which is actually quite a good way of looking at the worst possible outcome.

In this scenario we face a technological meltdown. Those few banks that are still standing after a run of customers demanding cash, lose their financial records including the assets of all their customers. Starved of raw materials and staff afraid to leave home, businesses go bust all over the place. The stock market crashes. Manufacturers find their factories won't work. People find themselves with no jobs and no money coming in. The government's computers go just as badly wrong, so they can't collect taxes which means they can't pay social security, or hospitals or local authorities. The food distribution system packs up. People start to go hungry and in their search for food, crime breaks out widely. Because the transportation network is disrupted, petrol and fuel supplies don't get through, the police and army find themselves immobilised, so anarchy takes over and Society grinds to a halt.

The doomsayers see us suffering severe power outages lasting for weeks or even months, possibly forever. Hospitals are unable to work with no power. Security systems trip out, opening the door for even more looting. Streets become alleys of darkness with feral pets on the prowl after being thrown out by owners who need the food for themselves. The electric locks on the gates of zoo animals trip out letting them add another frightening dimension.

Let your imagination run on this one and you'll soon start to realise that although they might be whacky and have their own agendas, the thin thread that holds our society together runs entirely through the computer network. In fact a report on the situation showed that by the spring of 1999 one in three Americans in the computer industry had become convinced that the start of 2000 will not be marked so much by celebration as civil unrest and rioting.

The believers in TEOTWAWKI also like to remind us

about the huge tracts of the world where there is simply no money allocated to millennium compliance. They describe the former Soviet Union as 'Bangladesh with Nukes,' and point out that Russia does not have enough money to feed its people, let alone invest in high-flying computer consultants. They almost seem to relish the suggestion that on 1/1/00 russia's vast arsenal of nuclear weapons might take to the skies upon discovering that they have apparently ceased to exist.

We can only hope that a side effect of glasnost was to take them off-line, because at the height of the cold war both sides boasted that their ICBMs were set to fire themselves if their control infrastructures were taken out. From the computer's point of view, they were not given the chance to philosophise – Any problems and it could be "Good Night America."

In fact I heard on the grapevine that the CIA has been spending some time with its Russian counterpart discussing this very problem, and indeed their worries about the Soviet nuclear submarine fleet currently tied up awaiting dismantling. One rumour doing the rounds is that the group's first move was to set up a guaranteed hotline that could be used to avoid an accidental launch of a Russian missile into an all out war as the US retaliates.

Whichever view you come to, there is no doubt that the doomsayers are having a heyday. They believe that 'The End of the World Is Nigh,' and quote at length from Revelations, Nostradamus and the Mayan Prophecies to support their beliefs. Not since the Cuban Missile Crisis have so many people found a single event on which to focus their energies.

There is a delicious irony in the fact that all these bodies of concerned people, convinced that computers are society's downfall, should promote their views to the world through the Internet.

The worrying thing is that they could be right. If the

power goes down for any length of time, then it won't matter how well compliance programmes have been run, all bets will be off.

THE MOST LIKELY OUTCOME

The most likely outcome lies somewhere between 'Business as Usual' and 'TEOTWAWKI.' The problem is that absolutely nobody can say where. There is no historical data to support any prediction and nobody will know for certain whether or not their systems will work until the event itself.

One of the problems with the Millennium, Bug is that it is a world-wide issue that will hit the whole world in the same 24 hour period. The world's airline industry flies on Greenwich Mean Time, so it will be the only sector that is hit all at once. For the rest of us there will be a sweeping effect as Midnight arrives first in New Zealand then progresses around the world to get back there 24 hours later. It could be the most devastating 24 hours in the history of civilisation, and it is a systemic problem that will affect everybody in the world. Not since the dinosaurs were wiped out, has the world faced a similar situation.

Or it could be the world's greatest party. Both scenarios are equally likely, but as with all things, the outcome will not be black and white but varying shades of grey.

Whatever happens, the West should at least have some forewarning, because the news media and interested individuals are setting up millennium bug watches to report what effects the bug causes as it sweeps the world,

Guess what? – Their reports will be available on the Internet. Assuming of course it is still there!

How do you balance the risk and the cost?

If you do decide to prepare for some form of millennium

chaos you have to make a decision as to what level you wish to go to. To a great extent that decision will be financial.

If you believe that nothing is going to go wrong, then that's how much you'll prepare. If you're right then congratulations. If you're wrong you could face severe privation, possibly worse.

If you believe in TEOTWAWKI, as do a lot of people in the US, then the received wisdom is that you'll sell everything except basic survival stuff, take all your money out of the bank and put it into things that you will be able to barter in a bank-free society, pack up your family head to the hills and live like a jungle soldier.

If you do all that and January 2000 passes by uneventfully, you will feel a total fool. You'll have made such a major life change that you'll never be able to restore your lifestyle of today, and will more than likely end up as the laughing stock of everybody who knows you.

If on the other hand your prediction comes true, you'll be the one with a smile on your face.

If the outcome lies in the middle you'll still be ahead. You'll certainly out-survive the people who were totally unprepared, but as society starts to pick itself up again, you'll still be significantly out of pocket.

Have you noticed that so far every paragraph in this section starts with 'If'? It just confirms that the whole Millennium Bug outcome is speculation. They've always said that you need to speculate to accumulate. This has never been more true.

The dangers of being wrongly-prepared

Let's look for a moment at the downsides of being wrongly prepared.

In the 'you wish' scenario you don't prepare at all. It doesn't cost you anything this year, but it could cost you dearly in the next. Even if disruption only lasts for a few weeks, you will find yourself going hungry, cold and, worst of all, thirsty. Even if the infrastructure survives unscathed, the very least that is likely to happen is a serious recession caused by companies and some countries going bust as a result of not being millennium-compliant.

If, like me, you're carrying a bit too much middle-aged spread, then you should last a week or three, provided you have water and a warm duvet. It might even help to achieve the svelte figure you've always promised you'd get back!

But if the disruption lasts for four or five weeks, or three or four months, then you'll soon stop congratulating yourself on your new shape and start to worry. For sure, there will be a lot of people in the same boat, but we no longer live in a society that pulls together, so you can pretty well guarantee you'll be on your own. Unless you're fortunate enough to have a close family member who has rations to spare, your diet could end up a lot more effective than you think.

If the TEOTWAWKI outcome hits and you're totally unprepared, then you might as well pack up and wait to die – you'll have no food, water or warmth and there will be no infrastructure in place to support you. You could fall ill and find there would be no hospitals or drugs you could get to, and the outcome of that could be terminal.

The downside of being ill-prepared is severe discomfort, possibly death, whereas the downside of being over-prepared is financial. What you have to do is to balance your perceived risk with your financial ability to mitigate it.

Nobody can look into a crystal ball and tell you what will happen, any more than they can tell you that you will leave home and drive to your destination without having a problem on the way. They can predict the likelihood of a car crash, or a mechanical breakdown, because there is 50 years of historical data on which to base their predictions. There is no historical data on which we can predict the outcome of the Millennium Bug. All you can do is to weigh up the facts at your disposal, listen to those who are in the know, and are prepared to talk about it, and make your own decision.

The survivors will be those who get that decision right. Personally I think that TEOTWAWKI does not have to be the outcome. If society prepares itself for the disruption, without destroying itself in the process, then we should get over the hump. Bear in mind that although the rate of change in society has been on an exponential curve for the last forty years since we started with computers, mankind has been on the planet for at least four million years beforehand, and did a pretty good job of surviving.

Pulling together

If there has been one common recurring thread in popular writing over the last three or four decades, it has been the 'Armageddon Story.' Let's be honest, nothing sells papers like bad news, and the thought that society could be facing a re-run of the Asteroid versus Dinosaur scenario does hold a gruesome fascination for just about everyone.

But how many people have really thought about what it would be like to live in a post- apocalyptic world? Everything that we now take for granted, from food in shops to central heating, ready medicine and a welfare system would cease. Money would lose its value, as would just about anything that could not be eaten or burnt.

How well would you survive? Probably the honest answer is not very. Even if you lasted for a few months, through living frugally off your supplies and keeping out of the way of trouble, then you would still have to accept the reality that one day you could no longer do it on your own.

There are many people who are hailing the potential catastrophe that the Millennium Meltdown represents as being not a catastrophe at all, but the opportunity for mankind to go back to basics and start again. They cite the problems of global warming, the exploitation of the environment, the introduction of genetically modified foods, and the way that society has moved from a co-operative effort by people to help each other, to being a combative place where success is measured purely in terms of material wealth.

In their view, society will be forced to pull together. They are already suggesting that communities should work together – That towns, villages, streets, even just neighbours, should discuss the potential for catastrophe and co-operate to achieve a common plan that will see them through.

There's quite a lot of sense in their idea.

I wrote this book because I believe profoundly, as does just about everybody who has helped in its writing and research, that whatever happens there will be significant disruption to life as we now take it for granted. Many forecasters are suggesting a deep global recession at the very least, while others are far more pessimistic,

Fundamentally what the pessimists forget is that man is an incredibly resourceful and creative animal. I mentioned in the first chapter that the Millennium Bug looks like it could be the ultimate key to Pandora's box. Remember that after all the troubles had flown the box, the last thing left inside was Hope.

Even if the Bug does bring us to TEOTWAWKI there will be survivors who will eventually come together and rebuild society, perhaps in a more sustainable form.

Hopefully you will be one of them.

11
Checklists

The following pages contain a by-no-means exhaustive set of lists of things you should either obtain, or make sure you do well in advance of Millennium day.

The whole point of this book is to give you the basics that you can build on to get you and your family through the worst of the crisis. What you decide to do is up to you, but within the lists you should find that most things will come in handy somewhere in your life.

Hopefully you won't need to rely on any of this stuff for more than the normal fun of living, but if you do you will be grateful you had it.

If you don't, then think of it in the same way as a car insurance policy - you happily pay for that every year, but how often do you have to claim?

Remember the motto of the SAS

PREPARATION
PREVENTS
PANIC

Around the home

Carry out a millennium bug health check on your domestic electrical equipment by contacting the manufacturers, retailers or service agents. Don't try carrying out 'self-checks' because as they could cause you more problems that they will solve.

Equipment you should check includes

TV
Video Recorders (VCRs)
Home PC
Clock radio alarm
Fax Machines
Answer Phones
Digital Cameras
Camcorders
Videophones
Watches
Burglar Alarms
Security systems
Lifts (e.g. in a communal building)
Gas Meter
Electricity Meter
Central heating programmer & Thermostat
Controller for cooker
Controller for boiler

Things to acquire / make sure are working

Water carrier / storage
Drinking water carrier
Gas barbecue
Gas bottles
Clothes Airer
Sleeping bags
Camp beds
Torch / Lantern
Replacement batteries
Kerosene lamps and fuel
Portable camping gas lamps and canisters
Portable battery or wind-up radio
Matches
Cigarette lighter plus spare fuel / flints
Stock of newspapers for lighting fires
Ant Powder
Bug zapper
Rodent killer
Sturdy winterproof clothing for whole family
Waterproof suit
Overshoes / boots
Gloves
Umbrella
Candles
Torches
Batteries
Electrician's Adhesive tape
2 Lengths of plastic tube for siphoning water / fuel
Bow saw / Chain Saw and fuel for felling trees for
firewood
Basic fishing gear
Cash

Domestic cleaning / hygiene materials

Water Purifying tablets
Soap Powder
Liquid Soap / hand soap
Shampoo
Anti Nit treatment
Sterile wipes
Toilet paper
Toothpaste,
Deodorant
Stock of towels
Paper towels
Ladies' Sanitary towels
Bin Liners
Stock of plastic carrier bags
Chemical toilet solution
15 gallons of bleach - use for sterilising water as well

First Aid kit

A good first aid book
Thermometer
Three month's supply of daily prescription medication for all family members
Antibiotics
Ointments for the eye, fungus & cuts
Anti-diarrhoea medication
Pain and anti-inflammatory medication e.g. paracetamol, Ibuprofen
Non-steroidal anti-inflammatories
Burn treatment
Sun Cream
Anti sting cream
Iodine
Disinfectant , e.g. Dettol
Oral electrolytes (for dehydration from fever, diarrhoea, stress)
Cold / flu remedies
Bandages
Gauze
Cotton
Surgical tape
Scissors
Scalpel plus spare blades
Tweezers
Needles to remove slivers
Dental kit to patch dentures or replace fillings,

Things to do well in advance

Service the car and any powered equipment, e.g. chain saw, generator
Buy Jerrycans and store petrol
Lay in one year's worth of coal or logs
Lay in a stock of water. (55 gallons will last a family of four for fourteen days)
Rehearse family in how to use emergency equipment
Learn how to cook from basic ingredients
Replace batteries in all domestic equipment, e.g. radios, torches. clocks, smoke alarms
Buy and fit smoke and carbon monoxide detectors
Build and store food supply
Check all doors and windows in the house are secure
Develop relationships with local farmers and produce providers
Buy and read books on how to grow your own vegetables
Prepare a vegetable patch in your garden
Try your hand at cooking from basic ingredients
Make sure everybody in the family is up-to-date with injections, especially typhoid and tetanus
Practice your barbecuing skills
Install a rainwater collection system
Build and try a water filter system
Pay down any debts
Build a reserve of cash
Discuss action plans with friends / family

Basic tools you might find useful

Hammer
Phillips Screwdriver
Selection of flat screwdrivers including electrical
Electricity mains tester kit
Craft knife and spare blades
Chisel
Pliers
Set of sockets
Hand operated drill plus selection of bits
Sledge Hammer
Large and small wood saws
Hacksaw
Several pairs of sturdy working gloves
Mole wrenches (x2)
sharpening stone
Bicycle repair kit

Garden tools
Axe
Bow saw
Pruning saw
Set of shears
Lopping shears
Spade
Fork
Hoe
Wheelbarrow
Trowel

About the Author

James Carpenter is an investigative journalist who has written extensively about business, computers and the Millennium Bug over the last five years. His work appears regularly in national press and consumer magazines as well as local radio.

He is well qualified to comment on the situation as he also writes corporate literature for a number of global companies, including several computer manufacturers.